THE QFD HANDBOOK

THE QFD HANDBOOK

Jack B. ReVelle

Hughes Missile Systems Co.
Tucson, Arizona

John W. Moran

Changing Healthcare, Inc.
Littleton, Massachusetts

Charles A. Cox

The Compass Organization
Higley, Arizona

JOHN WILEY & SONS, INC.

New York / Chichester / Weinheim / Brisbane / Singapore / Toronto

Designations used by companies to distinguish their products are often
claimed as trademarks. In all instances where John Wiley & Sons, Inc. is
aware of a claim, the product names appear in initial capital or all capital
letters. Readers, however, should contact the appropriate companies for
more complete information regarding trademarks and registration.

This text is printed on acid-free paper.

Library of Congress Cataloging in Publication Data:
The QFD Handbook / Jack B. ReVelle . . . [et al.].
  p. cm.
  Includes index.
  ISBN 0-471-17381-9 (cloth : alk. paper)
  1. Quality function deployment. 2. Production management–Quality
control. 3. New products–Management. I. ReVelle, Jack B.
TS156.Q275 1997
658.5'62–dc21   97-6508
                                                                       CIP

Printed in the United States of America

10 9 8 7 6 5 4 3 2 1

To Our Wives, Brenda, Claudette, and
Marilyn; and to our progeny, Karen, Wally, and Kate & Chad—the
QFDers of the future.
And to our readers, may you all reside in houses of quality.
Jack R., Jack M., and Chuck

CONTENTS

FOREWORD

It was 1983 when I first introduced Quality Function Deployment (QFD) to the United States. Dissemination and development of QFD in America have been remarkable since then. In recent years, QFD in this country seems to have progressed further than in Japan.

The First International Symposium on QFD (ISQFD) was held in 1995 in Japan. The following year, the Second ISQFD was held on June 10–11 in the United States. It was attended by nearly 300 people from over 25 countries, with 36 case studies presented that were followed by lively discussions. The Third International Symposium on QFD is planned for Sweden on October 1–2, 1997, and the Fourth in Australia in 1998.

Publication of *The QFD Handbook* by John Wiley & Sons in the midst of this QFD growth will benefit not only the companies planning to introduce QFD in the future but also those that already have begun the process, as the book will help them make future progress.

This book contains not only the basics of QFD but also Taguchi Methods and Failure Mode Effects Analysis (FMEA) and use of the recently introduced Theory of Inventive Problem Solving (TRIZ), Business Process Reengineering and other methodologies in QFD. Furthermore, this book is unique in that it also discusses applications of QFD in the quality systems ISO 9000, QS 9000, environmental product life cycle and other leading-edge areas. Managing and monitoring QFD, in particular, will be useful as a guide when a company actually becomes acquainted with QFD.

It was around 1966 that I first began conceptualizing and testing QFD, without getting much attention at that time. What today is called QFD was formulated from my 1972 publication on this subject and the subsequent quality table proposed by Mitsubishi Heavy Industry's Kobe Shipyard. After experiencing a period of growth in Japan, QFD was introduced in the United States, where its use became widespread. Now we are seeing worldwide dissemination taking place.

Publication of this book is timely. The authors studied QFD from its earliest inception and have actually applied QFD. It is my hope that QFD will see further dissemination and progress with publication of this book.

YOJI AKAO

ISQFD '96, Novi, Michigan

PREFACE

Congratulations! By acquiring this book, you have taken an important step in expanding your understanding and ability to make full use of Quality Function Deployment (QFD) in a broad variety of venues.

The authors have prepared this text to provide you, our readers, both experienced and neophytes in the practice of QFD, with a broad-based collection of highly relevant readings all of which are focused on the selection, application and management of QFD-oriented assignments.

The authors are themselves diversified in their educational and experiential backgrounds. We believe that this is a major advantage in that we, along with the many contributing authors, can bring our readers important information that will allow them to make better use of QFD. We are fully aware of the many excellent QFD books that already exist in the marketplace, but with few exceptions, they are all single-author texts. For this and a variety of other relevant reasons, we chose to combine our various resources so as to produce what we believe will be a major treatise on QFD.

The authors take this opportunity to thank the contributing authors, whose chapters and appendices have added immeasurably to the value of this text. Without their valuable insights, the scope of this handbook would be much narrower. These individuals were asked to contribute to this text because of their reputation as leaders in specific application areas.

This text attempts to offer information in logical "chunks" so as to make its use by our readers easier and more relevant to their current challenges. The appendices are provided as a supplemental source of QFD information that, in several cases, goes well beyond the scope of the text.

A feature of *The QFD Handbook* that will make it unique in your library of QFD resources is that it includes a copy of the QFD/Pathway software. This software is unlike any other QFD software in that it provides QFD users with an opportunity to plan their QFD activities in advance of initiating their use of QFD.

QFD/Pathway is a new, user-friendly, expert system designed by experienced QFD practitioners. It provides necessary assistance in moving a QFD activity from the House of Quality (HOQ, or A1 matrix) to the next logical steps. This includes one or more matrices that are graphically identified in the GOAL/QPC Matrix of Matrices as well as other more recently developed matrices. QFD/Pathway assists a team to move from translation of its goals, resources and strategies to the selection and sequential deployment of specific QFD matrices.

The typical user of QFD/Pathway is most likely a team of QFD practitioners with a blend of QFD experiences. Whatever form of documentation may be currently employed, we have found that the typical team becomes stymied due to uncertainty regarding where to go from the HOQ. This is eliminated through consistent use of QFD/Pathway from the outset and as needed as long as the team exists.

We suggest that your first action following acquisition of this text should be to thoroughly review the table of contents to ensure that you are fully aware of its scope and depth. Then, we encourage you to read a few of the chapters and/or appendices that are especially pertinent to your current areas of interest. You may even want to let your colleagues know that you have a copy of *The QFD Handbook* with the QFD/Pathway software and that you would like to share these with them.

QFD was created to help organizations improve their ability to understand their customers' needs as well as to effectively respond to those needs. For us to make the revised edition of this text better than this first edition, we need you to let us know what you think of our product. Give us your "verbatims" and we will deploy them into our revised edition. We can be contacted through the publisher, John Wiley & Sons, Inc., 605 Third Avenue, New York, NY 10158.

JACK B. REVELLE, Ph.D.
JOHN W. MORAN, Ph.D.
CHARLES A. COX

ACKNOWLEDGMENTS

The authors would like to thank those who contributed to this effort. Some of the graphics and early typing were provided by John W. Moran III, a student at Trinity College.

We also recognize Marilyn Cox for her typing, proofing and editing.

Lastly the authors extend their thanks to the following individuals for their particular contributions, critiques and support in the development of this text: Robert F. Hales, Glenn H. Mazur and Daniel F. Nissly.

CONTRIBUTORS

David Anderson, Anderson Consulting and Seminars, P. O. Box 1082, Lafayette, CA 94549-1082

William Harral, Arch Associates, 15770 Robinwood Dr., Northville, MI 48167

Kurt Hofmeister, Total Quality Group, 13061 Glenview Rd., Plymouth, MI 48170

Glenn Mazur, Japan Business Consultants, Ltd., 1140 Moorhead Court, Ann Arbor, MI 48103

Henry Rogers, 16651 Canyon View Dr., Riverside, CA 92504

Shin Taguchi, American Supplier Institute, 17333 Federal Blvd., Suite 220, Allen Park, MI 48101

John Terninko, Responsible Management, Inc., 62 Case Rd., Nottingham, NH 03292

Steve Ungvari, Strategic Product Innovation, Inc., 7591 Brighton Rd., Brighton, MI 48116

David Verduyn, American Supplier Institute, 17333 Federal Blvd., Suite 220, Allen Park, MI 48101

Alan Wu, American Supplier Institute, 17333 Federal Blvd., Suite 220, Allen Park, MI 48101

Richard Zultner, Zultner & Company, 12 Wallingford Dr., Princeton, NJ 08540

THE QFD HANDBOOK

I

INTRODUCTION TO QFD

1

QFD: ITS ORIGINS
AND OBJECTIVES

BACKGROUND

The need for Quality Function Deployment (QFD) was driven by two related objectives. These objectives started with the users (or customers) of a product and ended with its producers. The two objectives were

1. To convert the users' needs (or customers' demands) for product benefits into substitute quality characteristics at the design stage
2. To deploy the substitute quality characteristics identified at the design stage to the production activities, thereby establishing the necessary control points and check points *prior* to production start-up

If these two objectives were met, the result would be a product designed *and* produced that met the users' needs and customers' demands for product benefits.

At the outset the Quality Table was the tool used to assist in the conversion of customers' demands to substitute quality characteristics. Its matrix structure and visual nature gave both discipline and guidance to the conversion process. Follow-on matrices communicated (deployed) the substitute quality characteristics to production operations. This approach is known as Quality Deployment.

The basis of the current QFD-style matrices, originally referred to as Quality Tables, was first proposed and used by Mitsubishi Heavy

Industry's Kobe Shipyards to design supertankers. In May of 1972 Mitsubishi Heavy Industry published a discussion of the Quality Table in a magazine article. (The pioneering concepts and activities of Mizuno, Akao, Furukawa, and others that led to these first Quality Tables are covered in Appendix E, Short History of QFD in Japan.) Although the Quality Table is important to QFD, it is just a matrix. Matrices have been used for a wide variety of purposes in the United States and elsewhere since the 1950s. For example, economists have used matrices for input–output analyses and building models of entire economies. Matrices have long been used as a tool to understand relationships between all types of inputs and outputs, but they alone could not have produced the integrated outcome of what is now called Quality Function Deployment.

The concept of quality deployment was first proposed by Akao in 1966 and expanded upon in an article published in 1969. Akao published the idea as a system in an April 1972 magazine article under the name Hinshitsu Tenkai System. (He was the first to use the term *hinshitsu tenkai*, or quality deployment.) The publication in 1972, in separate magazines, of Akao's Quality Deployment and Mitsubishi Heavy Industry's Quality Table was followed in 1976 by Akao creating the system that later became known as the QC Process Table. In 1978, Shigeru Mizuno, together with Akao, published the very first book on QFD (translated in 1994 under the title *QFD: The Customer-Driven Approach to Quality Planning and Development*).

Toyota Auto Body had developed a quality table that had a "roof" on top, and in fact, by then, Sawada of Toyota Auto Body was already using the term *quality house* (which was presented during a Japan Standards Association conference in 1979). This was passed on through the American Supplier Institute (ASI) by Fukuhara, who was originally with Toyota Auto Body, as the *House of Quality,* which became the nickname of the Quality Table (see Figure 1-1). (Note that the Toyota Auto Body Company was the earliest to adopt the QFD methodology, not Toyota. It was not until 1979 that the entire Toyota Group began introducing QFD.)

QFD was formally introduced to the United States in 1983 by Furukawa, Kogure, and Akao during a four-day seminar for about 80 quality assurance managers from prominent U.S. companies. Also instrumental in the introduction of QFD in the United States was the article by Kogure and Akao in the October 1983 issue of Quality Progress, "Quality Function Deployment and CWQC in Japan."

Subsequent development and refinements at Toyota Auto Body in the mid-1970s led to interest on the part of Ford. In 1984, Donald

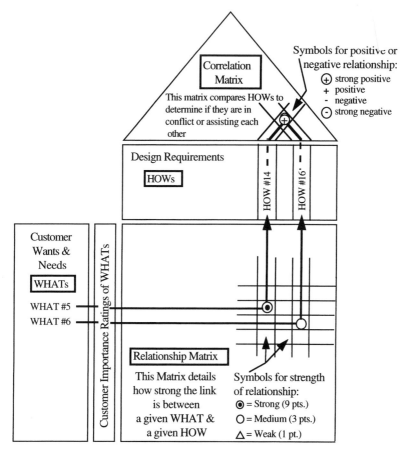

Figure 1-1 The House of Quality.

Clausing of Xerox introduced QFD to Ford. A year later Larry Sullivan (founder of the ASI) and John McHugh set up a QFD project with Ford Body and Assembly and its suppliers. In 1987, The Budd Company and Kelsey-Hayes, both Ford suppliers, developed the first case study on QFD outside Japan. One of the authors' first exposure to QFD was as a quality manager for Kelsey-Hayes during that study. Also in 1987, Bob King's (1987b) book, *Better Designs in Half the Time: Implementing QFD in America,* was published. Soon thereafter, John Hauser and Don Clausing published their article, "The House of Quality," in the May-June 1988 edition of the *Harvard Business Review.*

Since that time QFD has been used in ever widening circles. Besides Ford, General Motors, Chrysler, AT&T, Bell Labs, USWest, Polaroid,

Digital Equipment Corp., Deere & Company, Proctor & Gamble, Scott Paper, Ritz-Carlton, DuPont, Dow Corning, Eastman Kodak, Texas Instruments, Hewlett-Packard, General Electric, Allied-Signal, McDonnell Douglas, Hughes Aircraft, Martin Marietta, Rockwell International and Florida Power & Light also use QFD in a variety of applications.

Although the QFD methodology is widely used, the term "Quality Function Deployment" is awkward and does not immediately give an understanding of what the methodology is or does. It comes from the original Japanese phrase consisting of three characters, each of which has several meanings:

Hin shitsu, which can mean "quality," "features," "attributes," or "qualities"

Kino, which can mean "functions" or "mechanisms"

Ten kai, which can mean "deployment," "evolution," "diffusion," or "development"

Basically QFD means deploying the attributes of a product or service desired by the customer throughout all the appropriate functional components of an organization. QFD also provides a mechanism for its achievement, that is, the set of matrices that serves as both a structure and a graphic of the deployment process. In fact, there are three similar but different approaches to QFD that were developed by Akao and Makabe and Fukuhara. Akao's approach is the basis of the GOAL/QPC Matrix of Matrices developed by Bob King, which includes some 30 matrices. Makabe's model was introduced to Ford by Donald Clausing and became the basis of the ASI Four-Matrix approach. Fukuhara provided QFD leadership at Toyota Auto Body where 18 matrices were used in the now classic rust studies. A greater understanding of the application of the different approaches is developed in Chapter 23, Alternative Approaches to QFD.

Many people refer to QFD as yet another quality tool. It is, in fact, neither a tool nor solely associated with quality. QFD is a product/service/process/software planning process. Doing the planning process well cannot help but deliver a higher quality result to the customer, but what QFD is all about is taking the customers' inputs and translating them into technical specifics that engineers and other knowledgeable persons can act upon. Then QFD establishes a means to measure and monitor how well the technical specifics have been accomplished. Using the QFD methodology, if there are strong linkages between the customers' inputs and the technical specifics, then high performance on the technical specifics will result in superior performance from the customers' perspective. In developing a new product, service or process, there may not be a complete set of well-defined specifications. In fact, it is often the case that some of the customers' key performance characteristics are not mentioned at all, whereas other characteristics, which are known, are of little importance to the customers!

However, it is possible using QFD methodology to overcome this lack of definition and still deliver a great product or service. This is accomplished by a disciplined gathering and analysis of the Voice of the Customer. The foundation of the QFD approach is the information on the customers' wants, needs and demands as well as the customers' prioritization of these desires. QFD is based on the information gathered from actual customers/users of the product, service or process; their statements on their needs, wants and demands; and their prioritization of these desires. The entire QFD project and all of the design team's activities are driven by the information gleaned from actual users and customers. Besides this all-encompassing focus on the customer, several other aspects should condition a person's view of the QFD methodology.

First QFD is not new. The QFD approach and associated matrices have been used for over a quarter of a century. QFD is, in a way, advanced proactive cause-and-effect analysis as well as a form of quality assurance. Once American auto companies saw the results that Japanese auto companies were getting using QFD, they too began using it and adapting it for their needs, mostly for designing products. As QFD became more widely known, people began using it to design not only products but also services.

QFD is not difficult to use. It is an effective way of taking input from users and customers (ideas, concepts, ways the customer utilizes the product or service or process) and using those inputs to create very successful outputs that are highly focused and responsive to the customers' inputs. Although many firms use QFD matrices for designing products or services, some firms have used it to create their strategic plan. Others use it to design a completely new process when they are involved in business process re-engineering. Anywhere there is a need to create an effective design, QFD is an effective and efficient approach.

Second, QFD is an awkward name for a wonderfully well thought out, integrated (synergistic) set of activities. What is not awkward, and the basis of QFD's importance to any organization that has a design function (whether it is for new products, services or processes), is how QFD melds three powerful concepts into a single design process model:

(A) Transition from Customers' Jargon to Technical Specifics. QFD provides a structured transition from the customer's general, sometimes not well-defined, expressed needs and wants to the technical specifics that engineers and technicians must possess if they are to have a real opportunity to achieve a well-designed product/service/process. And, using the customers' stated importance rankings, it is possible for the design team to organize and allocate their limited resources toward the goal of maximizing their impact on the customers' most important wants.

(B) Rational Representations of Linkages between the Customer and the Design. QFD uses a representation of the transitions (linkages) that is easy for a person/team to relate to and understand because it is both graphical and rationally structured to demonstrate all of the linkages. For example, in its final form, a typical House of Quality may seem like a busy set of matrices, but the team that gradually built that House of Quality, area by area ("room by room"), can easily walk a non–team member through it and explain the background for each set of decisions and how each room links to the rest of the House. (See Figure 1-1.)

(C) Knowledge Gained from a Multifunctional, Interactive Design Team. The more diverse, knowledgeable and interactive the QFD design team is, the more robust the resulting design. QFD design team members are often colocated to assist an ongoing exchange of information and to increase the overall exchange of information between the members. This proximity of the team members plus their varied backgrounds encourages an integrated approach to the design in which no factor is overplayed on the one hand and no key factor is ignored on the other.

COMMON ASPECTS OF QFD PROJECTS

Integration of Information from Multiple Sources

The information needed to create the various matrices, especially the first matrix (called the House of Quality, or A1 matrix), comes from a wide variety of sources. Acquiring the information typically requires inputs from marketing sources (internal and external), technical sources (internal and external, in some instances including suppliers of some state-of-the-art components), field service personnel, customer service representatives and, most importantly, representatives from several different sets of key customers. Most (or all) of these groups should be represented on the QFD design team. The larger the area over which the information "net" is cast, the more likely the end result from the QFD design process will fully meet (or exceed) the customers' expectations.

Maintenance of Well-Structured Stages

The earliest activities in the design process are well structured, saving waste and (costly) iterations later. Using an approach such as QFD is so important because in the very earliest stages of any design effort many decisions are made that impact the final results. Just deciding what concept will be worked on and developed further defines and commits about 80% of the final costs (processing and assembly labor, raw materials, processing and purchased parts, for example, in the case of a manufactured product). It also decides the total cycle time (product or service), the durability of the product, the ease of consistently delivering the service or product (one measure of quality) and several of the customers' perceptions of quality of the product service or process. The QFD methodology starts with collecting information on the needs and wants of the customer. However, just knowing the stated needs and wants does not give enough understanding. It is also important that the customers' perceptions of quality are well understood. Knowing how the customer perceives quality is very important when the concept(s) that the product or service are going to be based on are chosen. For example, a product that uses steel components might be viewed by an average customer as being stronger and more durable than a similar product made from plastic components that have been glued together. This could occur even though the plastics and the "glue" might really amount to a composite material that could prove to be stronger and more durable than the welded steel. The customers' perceptions should always be considered as a part of the design process.

Here's an example of *kansei,* or sensory engineering. Consider the use of an artificial leather, grain-embossed vinyl surface that is applied on an auto dashboard in lieu of a plain, flat vinyl. This can work well for some applications when customers only look at the surface. However, it is not advisable to use the embossed vinyl as a steering wheel cover because the constant contact will smooth out the vinyl and, although vinyl may look like leather (at least initially!), it does not feel like leather. So every time the customer grabs the steering wheel he or she is reminded of the initial false perception of the "leather"-wrapped steering wheel.

"As the twig is bent, so grows the tree." This saying, usually meant to apply to children as they grow to adulthood, is also true for product concepts brought to life as products in full production, or freshly introduced services or newly designed and implemented processes. Earliest inputs must be "right" because they determine the direction of so many activities that follow.

Focus on Subsequent Steps

Another reason for using QFD is that it keeps the process of changing customer desires/wants/needs into technical specifications focused. There is much less wasted effort. The matrices capture, organize and display the information as it becomes known. Where the matrices have the information filled in, it indicates that at least the first iteration has been completed. Where there is no information displayed, it clearly indicates where the team has work to do.

SUMMARY

Why would you want to implement (or at least try) this methodology? The QFD approach allows the team and manager to understand the design process from beginning to end, including what resources will be needed, to what level and at what point in time. It is not a perfect map, but in application after application it has proven to be better, sometimes dramatically better, than any other approach tried. Because the design process is more carefully described, there are substantially fewer false starts. The outcomes resulting from the QFD design process are much more closely aligned to the needs and desires of the customer, so there are substantially fewer changes needed (fewer iterations to get it "right," fewer Engineering Change Orders, etc.) after the introduction of the product or service (or new process). Going through the actual design exercise creates a team that knows how to leverage other people's knowledge, experience and skills (i.e., their team members) so as to create better results.

In conclusion, some benefits of QFD are as follows:

1. It will give you a better product, service or process than you would have achieved otherwise.
2. It will give you this better outcome faster than will other methods.
3. It will typically require fewer resources.

4. It will give definition to the design process, helping the design team to stay focused and effective, giving team members greater ability to see and understand how they contribute to the design process as well as how to work with customers and other team members.

5. It will allow for easy management and peer review of design activities as they progress, with graphical representation of the different sets of information driving the design as well as the linkages between information sets.

6. It will leave you very well positioned should you need to improve upon your results for the next-generation product, service or process.

Besides these few benefits, it has nothing to recommend it!!!

2

QFD RESULTS

If a QFD approach includes all of the vital elements, such as cross-functional design teams, seeking the Voice of the Customer activities, organizing and prioritizing the design activities according to the information gained from the customer and so forth, there are many important, positive results that accrue to the different parts of the organization. One of the most important is the shorter time-to-market advantage that using QFD gives. With the new product well defined by the first QFD project, the history and knowledge that the first project's QFD matrices impart to the design team that performs the next-generation design give the organization a virtually unrecoverable lead over any competitor's design team. The results of using QFD impact everyone and everything. This is true from the overall organization's ability to compete down to the personal development of the team members and the functions involved in the design process.

In the 1994 Boston University Manufacturing Roundtable's U.S. Manufacturing Futures Survey (which surveys senior manufacturing executives from over 200 American manufacturers every two years) reported the following results: "The competitive challenge is to reduce price and enhance new product introduction speed with no sacrifice in basic quality and delivery." Survey participants reported that "the product development process and the order fulfillment process, both of which involve more than the manufacturing function, are becoming more critical than the conventional materials transformation process in meeting future challenges. These are the areas where larger opportunities exist for improvement in cost, time and quality. The alarming news is that the product development process has been improving at a much slower rate than other areas." The Roundtable's last three reports show that new product introduction has risen from ninth in competitive priority in 1990, to eighth in 1992, to sixth in 1994 [behind (1) conformance quality, (2) on-time delivery, (3) product reliability, (4) low price, and (5) fast delivery].

Fortune magazine (February 13, 1989) noted that a model developed by McKinsey & Co. shows high-technology products introduced to the market six months late but on budget earn 33% less profit over five years. However, introducing the product on time but 50% over budget only reduces profits by 4%. Gupta and Wileman (1990) surveyed managers in 12 large technology-based firms. Eighty-seven percent of the respondents reported that most of the reasons that delayed product development in the past continue to exist in their companies.

When asked the reasons for product development delays, 71% responded poor definition of product requirements, 58% technological uncertainty, 42% lack of senior management support, 42% lack of resources and 29% poor project management. The QFD approach helps dramatically to improve the definition of product requirements because it solicits and translates the customers' statements of need (the Voice of the Customer). QFD supports the management of the product design process by providing a defined process that shows what to do at each step. QFD can mitigate somewhat the lack of resources because it uses resources more effectively. The graphical approach of QFD increases the ease of communication and results in greater support through greater understanding.

Design team members gain insight from fellow team members both on a technical basis and in the team-building and joint decision-making processes. They learn how decisions made during the design phase affect the related downstream activities of manufacturing, process and tool design as well as both the distribution and service channels and the end user. All of the downstream activities that occur inside the organization are the "internal customers" to the design project. The people outside the organization are "external customers." When organizing to fulfill the expectations of your external customers, it is easy to focus exclusively on the people who buy the product or service. But if you truly expect to satisfy your external customers on a long-term basis, it is essential that all of the organization's activities identify and fill the needs of the internal customers as well. It is only by thoroughly serving your internal customers that you can address the needs of your external customers.

By having a cross-functional design team, the needs of both the external and internal customers are identified and are worked on simultaneously. Thus, the final product design has an excellent chance, once it is released, of going to fully ramped production with few, if any, Engineering Change Orders. Furthermore, product returns from the field due to inadequate design are eliminated. By simultaneously addressing internal and external needs, the result is a product that is easier to make and less costly to get into production. The QFD approach provides a much greater ability to communicate as well as to understand and resolve the different demands on various limited resources and to gain the needed experience and knowledge at the right time in the design process.

Senior management gets an up-front view of the difficulty of the design effort and will know many, if not most, of its elements. They will also know the reasons for making extra efforts (more time, money, expertise) to give the customer certain benefits. The initial matrix will not be able to offer complete intimate details but will certainly give strong indications what technology resources will be needed and the levels of effort needed from each resource. Some estimations of the time to reach design release based on the accumulated difficulties for each resource area can be made. From the various functions represented on the design team, middle management can understand what needs to be provided by their function to make the product a success. Front-line workers and supervisors can be briefed on the customers' requirements and how their activities relate to meeting and exceeding the customers' expectations.

QFD has also helped produce the following specific results:

- Customers of Motorola America's Parts Division were 60% more satisfied with their product and pricing information after the system was improved using QFD principles. The team said: "of several research methods considered, only QFD could satisfy all our needs. As we have learned, it is most critical to have a detailed understanding obtained directly from the customer." And now, they are going on to improve the customer order entry process. This focus on improvement is important because Motorola's customers ranked aftermarket support as second only to product quality.

- A Brazilian engineering company integrated German technology and equipment with some local erection contractors to install a multicompartment silo at a cement factory. Project development time was reduced by 20 days and construction time by 65 days while costs were reduced with innovative construction solutions.

- Toyota Auto Body Company in Japan reported a cumulative 61% reduction in startup costs related to the introduction of four different van models over a seven-year period. During the same time period the product development cycle was reduced by one-third.

- An architectural design and construction firm in Finland used QFD to check its design of a restaurant in an office building. Some of the changes that were made are the coffee dispensing area was downsized and oriented differently, more electrical outlets were installed in the dining room floor, storage spaces were redesigned, the dishwashing station soundproofing was improved, the kitchen layout was changed for better visibility and traffic patterns, cold storage facilities were rearranged, provisions for equipment to be added later were made and the service counter was altered and separated from the kitchen by a wall.

- A Brazilian steel company used QFD to reduce costs and increase market share in the rods and bars used in automotive suspension springs. Since 1993 its market share has increased by 120%, customer complaints have fallen by 90% and production cost is 23% lower and continues to decline.

- Host Marriott used QFD to study its sales of bagels in airports. Bagel sales 120 days after the QFD project were up 240%. That success has led to further studies of other foods and beverages using QFD.

- Dow Corning used QFD to determine employee job expectations and then to design the Employee Development Process, which impacts 10 of those expectations.

- The customers of Ritz-Carlton identified 19 critical processes as vital to a continuing business. Each Ritz-Carlton hotel championed one process. The Dearborn, Michigan, hotel took on the housekeeping system with the goals of 50% cycle time reduction, 100% customer retention and less than 3.4 defects per 1 million transactions. Using QFD the room-cleaning cycle was reduced 65%, defects per room were reduced by 42% with a 33% reduction in guest room interruptions. Overall housekeeping productivity was increased 15% and the distance traveled in the room was reduced 64%.

- Chevron USA uses QFD for curriculum planning and development in its employee training program. Management knows what resources will be needed to maintain what skills and employees know why they are taking each course.

- In 1991, The Wiremold Company applied QFD to its new product development process. By 1994 it had reduced new product development times by 75% (from 24–30 months to 6–9 months). It was able to introduce 16–18 products per year, compared to 2–3 products per year three years earlier. This was accomplished with no increase in salaried personnel. Simultaneously, higher quality products and dramatic sales growth were achieved.

- Nokia Home Electronics of Germany used QFD to develop a new generation of color televisions. The resulting design raised quality to a new level, and the total development time was nearly half of a similar previous project.

- Chrysler's Neon automotive design program used QFD to reach most of its targets for "Fun to Drive." (Even when they were not met, Chrysler came quite close.) For the key subjective targets, 5 to 6 Best in Class objectives were met and 11 of 12 competitive objectives were met.

- A ceramic tile manufacturer in Thailand shortened its standard manual system for product development from 13 days to 9 days (30%) with CAD systems identified as a source of further improvement.

- A hospital emergency room used QFD in conjunction with Taguchi Robust Design Methods to reduce the average patient length of stay 25% without any major capital expenditure. At the onset of the study an expansion costing several million dollars was proposed to handle the patient flow (the American College of Emergency Physicians guidelines suggest at least 11 beds, instead of the current 9 beds: 7 in examining rooms and 2 in the halls). If all of the new capacity is utilized at normal rates, additional revenue of $2.06 million is possible.

In 1995 the results were announced of a cross-sectional survey undertaken by the University of Michigan's Department of Industrial and Operations Engineering to understand the traits, attributes and general approaches to the usage of QFD in the United States. Respondents were asked to give their impression of a recent QFD project that is representative of QFD activities in their organization. This project, referred to as Project X in the survey, is the basis for collecting data on the factors associated with successful QFD. Table 2-1 summarizes the reported benefits of Project X QFD. Although the majority of respondents believed that the use of QFD had a positive impact on the project, over a third of the projects reported mixed results, no impact or that the QFD study was never completed. These mixed results are a reflection of the obstacles that have confronted QFD users. The primary problem reported by companies using QFD is their inexperience in using the methodology. It is expected that the other

Table 2-1   Project X QFD benefits

Project X Results	%	Product Improvement	%	Process Improvement	%
Significant impact	64.9	Increased customer satisfaction	82.7	Facilitated rational decisions	76.0
Positive impact	65.6				
Mixed results	19.8	Improved design	66.7	Increased corporate memory	73.7
No impact	4.2	Improved initial quality	52.6		
Aborted	10.4			Improved team unity	67.0
		Increased design alternatives	46.8	Improved marketing/ design communication	62.1
		Reduced number of changes	45.8		
		Increased innovation	44.9	Improved long-term cross-functional relationships	57.6
		Improved manufacturing	34.2		
		Increased sales	30.5	Improved design/ mfg. communication	51.2
				Reduced lead time	24.7

Source: University of Michigan, Department of Industry and Operations Survey on QFD Practices in U.S., 1995.

major difficulties identified (e.g., collection of information, matrices too large, judging the understanding of customer requirements, etc.) all share lack of experience as the root cause.

The primary product improvements are focused on customer satisfaction, which is consistent with the general customer focus of U.S. QFD. Only 34.2% of the projects reported that the Project X QFD improved the manufacturability of the product.

The use of QFD for Project X had a positive effect on the design and development process. Many intangible benefits of QFD were strongly valued by the respondents. These include improved decision making by facilitating rational decisions and improving communication and unity among team members. One noticeable contradiction with reported QFD benefits is that only 24% of respondents believed that QFD had reduced the lead time for Project X.

Key to implementing any product or process improvement strategy is organizational support. Since QFD is a planning tool, it results in a greater expenditure of time and money in the early phases of product development, when the opportunity to influence the design is at its greatest. The lack of organizational support in the form of money, time and top management support has been identified by previous case study research as being an inhibitor of QFD success. The results of the U.S. survey are consistent with this previous research.

The application of QFD transports the Voice of the Customer internally from the product and service design process to the factory floor or service site and externally into the marketplace. QFD has become the methodology of choice whenever there is a need to determine appropriate responses to the needs and expectations of customers. The benefits of QFD are both tangible and intangible. Some tangible benefits that you can expect to realize are

- Lower cost designs
- Elimination of most late engineering changes
- Early identification of high-risk areas
- Up-front determination of product process requirements
- Significant reduction in development time
- More efficient allocation of resources

Some of the intangible benefits of QFD are that it

- Increases customer satisfaction
- Facilitates multidisciplined teamwork
- Provides a basis for improvement planning
- Establishes and maintains documentation (the memory of team decisions)
- Creates a transferable storehouse of engineering knowledge
- Encourages transfer of training to other projects by the team members

QFD supports an organization's Total Quality Management efforts by

- Strengthening the current development process through
  - Early identification of goals based on user needs
  - Simultaneous focus on design and production
  - Highlighting and prioritizing key issues
  - Enhanced communication and teamwork

- Providing an objective definition of product or service quality through
  - Products and services that satisfy customer needs and expectations
  - Products and services that provide a competitive edge
- Facilitating team building and open communications because
  - Teams are enabled and empowered
  - Hidden agendas are eliminated

3

STEP-BY-STEP QFD: A STRATEGY FOR SUCCESS

In using the QFD methodology to design something, whether it is a product, a service or a process, the design team is expected to obtain and apply information not required by other design approaches. Just as modeling and market testing are necessary to answer certain questions, QFD activities have organizations formulate (and answer) new questions that they have not considered before. This approach may help an organization discover exciting and pertinent information previously overlooked. These discoveries are a crucial part of connecting the Voice of the Customer to the design. The knowledge gained gives the creators of the product/service/process a much better understanding of the needs of the user and the environment where it will be used (for brevity, the remainder of this chapter will refer to designing a product, although the approach applies to designing a service or a process). However, for QFD to be effective, it must be accepted as an integral part of the organization's design process. Every organization takes its own approach to designing, but all designing processes share some basic activities.

This chapter provides an overview of how QFD can fit into your organization's current design process as well as offers an alternative method for creating a design process. It is assumed that the goal of any design project is to move from a defined market segment (and a related product concept) to fully ramped production of that product as rapidly as possible using the optimal mix of resources and keeping the resulting product focused on the defined market. This must be accomplished with as few modifications to the design after release, often called Engineering Change Orders, as possible so as to meet and exceed the market's expectations for benefits from that product's features and functions.

SEQUENTIAL DESIGN

As shown in the flow chart in Figure 3-1, the sequential design approach tends to complete one activity before starting the next one. The first activity identifies the market or market segments that will be addressed by the design. Identifying the market for an existing product means defining the market you intend to capture and understanding its requirements. For a technology-driven organization, this involves matching features and functions of the new technology with the benefits that a

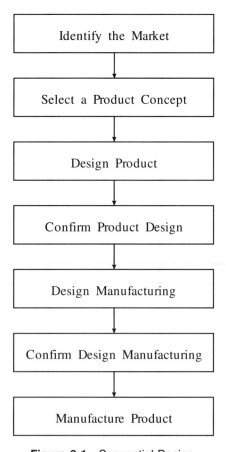

Figure 3-1 Sequential Design.

particular market expects. Such a technology-driven organization could also work to create a market for the product rather than introducing it into an already existing market. And, sometimes, a new technology "steals" market share from an older product or technology. For example, electronic mail and the Internet are taking over some of the market segments of fax machines and overnight courier services (a fax machine gets the text there immediately but it is not "letter perfect," an overnight courier service gets the hard copy or magnetic media there overnight and it is letter perfect, but a file attached to an Internet e-mail message gets letter-perfect text there immediately at a cost that is less than fax or overnight courier).

Once a business understands its market, it can begin the process of selecting product concepts. There should be up-front agreement on the criteria to be used to choose the best concept. Concepts tend to be much more successful (more easily manufactured, more easily delivered, more easily maintained, etc.) when factors downstream from the design process are taken into account during the concept selection process. Organizations involved with designing and manufacturing products joke about the design team tossing ideas over the wall separating it from manufacturing. The conclusion is that it is the manufacturing team's job to find a way to make a working version of the design. Having the manufacturing function involved early in the design activities (see Chapter 6, QFD and Designing for Manufacturability and Customization) is more effective because it allows a great deal of communication between the design and manufacturing functions.

Product design includes simulations, bench testing, research and development. The manufacturer may build prototypes and try them out with small, representative customer groups. (In the design of services or processes, simulations may be done for representative customer groups.) Designed experiments (see Chapter 7, QFD and Robust Design) may be necessary to optimize performance, minimize costs or both. Eventually, the team involved in the design of the product finalizes it and releases the design for production.

Depending on the relationship between the manufacturing processes and the new product's design (versus current products), changes in manufacturing methods may be necessary. Manufacturing may decide to build pilot lines or plants before ramping up to full production. Sometimes manufacturing in large volumes results in a different product than that obtained in pilot quantities, and adjustments need to be made. At some point, the manufacturing process is validated and standardized.

To the extent that the product design actually satisfies the customer and the manufactured product performs as intended, there will be no call backs from the field or engineering changes after fully ramped production commences.

CONCURRENT DESIGN

Concurrent design offers the opportunity to

(a) Compress the total time from market identification and concept selection to fully ramped production.
(b) Encourage better communications between different elements/ phases of the design process.

The staggered start of each of the activities requires additional communication across the functions involved in the design to ensure that each activity starts when it should, but not before the prerequisites have been satisfied. (See Figure 3-2.) To be successful using this approach, more structure is required. In addition, more discipline is required so that tasks or subtasks are not overlooked. The QFD methodology is a tremendous help because it breaks down the overall design process into many tasks so that it is possible to overlap the major activities. Also, the QFD structure assists the design team by guiding them through the sequence of tasks so none are overlooked. To accomplish concurrent design, the cross-functional team, which is an integral requirement of QFD, becomes an absolute necessity.

When QFD is being used on product design, a typical cross-functional team would include design engineering, process (manufacturing) engi-

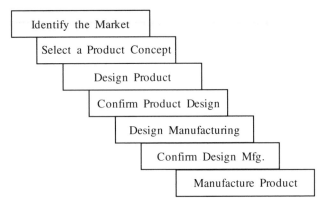

Figure 3-2 Concurrent Design.

neering, research and development (R&D), marketing/sales, quality assurance and test, field and customer service, manufacturing, in some cases distribution and, if the organization is project driven, project engineering. Note that even if an organization is not project oriented, especially those organizations that are trying to gain the maximum leverage from key suppliers' technology or are involved in coventures with either suppliers or customers, it is a good idea to have a person with project management skills leading and facilitating the design team.

PRODUCT DESIGN PROCESS: TIME AND RESOURCES

Persons reviewing the typical product design process who have had experience in designing many products often muse on what it would be like to (a) have a design process without the constant pressures of time deadlines and (b) have unlimited access to the best and brightest personnel in the organization as well as those with specialized knowledge outside the organization.

Would there be any differences between the actual design process the organization currently uses and a design process with no time or resource constraints? The answer is yes, definitely. First, with no time constraints, there would be no time line with milestones for progress reports. Second, the most talented people in the organization as well as outside suppliers and selected customers would have the time to serve as team members and maximize their contributions to the design process because they would not be distracted by other tasks and activities.

However, no matter how much we might wish otherwise, the actual design process is driven by the internal and external realities of our situation. The competitive marketplace and the availability and use of an organization's limited resources drive the product design process.

The marketplace places real time constraints on any actual design process. Time to market is almost always critical for market share either to gain market share or to maintain it. Some organizations predict lost sales as a function of time. Introducing a product today will yield the largest sales volume. In addition to the larger volumes, the earliest introducer can often obtain premium pricing on those volumes, at least for a while, and therefore produce a faster payback on the one-time expenses of designing, tooling and promotion associated with the new product. Each day that passes after a competitor has introduced a product reduces the market share for all the later entries.

Likewise, there is always a limit on the number of hours available for knowledgeable personnel to work on design issues. For the maximum output to be attained in the shortest period of time requires the design process to be well organized and structured with guidelines for both the internal, core design team members and the external members who provide some specialized knowledge. Using the QFD approach to design allows for much better utilization of limited personnel resources and allows the issues to be focused and worked on over a shorter duration.

THE MAJOR STEPS

This overview covers the five major steps in QFD. Step 1 (Figure 3-3) includes all the activities that focus on understanding the customer. The data produced are refined, and then a second subset of the information becomes the input for the second step. The prioritization of the customer segments is one of the outputs from step 1. The analysis of the customer starts with identifying the customer (market) segments and their characteristics.

Step 2 (Figure 3-4) involves gathering the Voice of the Customer and understanding the context in which the customer makes statements. Contextual information clarifies the customer's verbatim information. The purpose of this activity is to establish a clear understanding of all of the customer's needs, particularly the subjective performance requirements. These subjective performance requirements are sometimes referred to as demanded qualities. Each QFD analysis builds in some way on these demanded qualities. With the design driven by customer information, the QFD design team is more likely to design a product that meets or exceeds the customer's expectations.

Figure 3-3 Step one: understanding the customer.

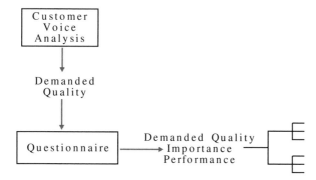

Figure 3-4   Step two: capturing and analyzing the voices.

The demanded qualities become the foundation of further activities, such as questionnaires, one-on-one interviews, focus groups, and so on, for gathering more information about how the customer will rank the importance of the various demanded qualities and the current level of customer satisfaction with them.

The matrix used in step 3 (Figure 3-5) translates the customer's statements and evaluations into the design team's performance measures, language and priorities. This translation is especially important because it typically takes the customer's language (such as "comfort-

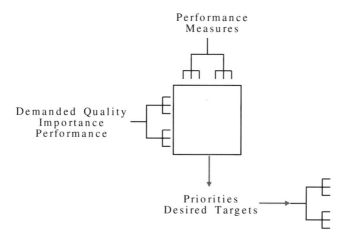

Figure 3-5   Step three: translating demanded quality into performance measures.

able," "user friendly" and "easy to start") and turns it into technical language with which the designers can work. The demanded qualities are the input to the matrix. In some instances, the design team may be able to use demanded qualities to identify new concepts that represent opportunities to be exploited to gain competitive advantage.

The design team sets priorities for demanded qualities by combining organizational priorities with the customer's priorities (importance rankings). The team also transforms the customer's subjective demanded qualities into technical performance measures that become the matrix output. The team uses these performance measures to prioritize the project's resource allocations, for example, which performance measures should receive what resources, so that in the end the design project's limited resources will have been allocated in such a way as to create a product with the greatest impact on the customer's demanded qualities. The performance measures are also used to establish desired target values for the design. These targets form the "wish list" that drives the design and development effort. Some compromises may be required when all of the wish list items are not attainable. It is important to note here that the design team is now working with much more technical language than the original subjective language of the customer. If the translation from subjective to objective has been done well, the technicians and engineers now have a much better chance of hitting their technical targets.

It is possible, sometimes even desirable, to add a few more rows to the bottom of the matrix. These rows will indicate (a) how difficult it will be to get to the target performance measure, (b) how long the development effort will take to reach the target and (c) whether the team is going to pursue the target (improvement) or merely the current specification (parity).

Step 4 (Figure 3-6) employs Stuart Pugh's system for generating and comparing new concepts. Target costs are often integrated into the generation process. The output of the previous matrix analysis becomes the input to this matrix analysis. The selected best concept and associated specifications are linked to the manufacturing process and the manufacturing database.

In step 5 (Figure 3-7), the final analysis uses a matrix to link the product specifications to the manufacturing conditions. Identifying the knowledge base for the relationship between operating conditions and product performance is part of the manufacturing database. The output of this analysis can be quality control systems or procedures for assuring that the manufacturing process is consistent with low variability.

QFD does not replace an organization's existing design process. Accordingly, you can integrate QFD into a sequential design process,

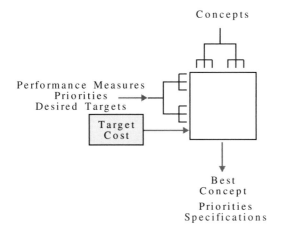

Figure 3-6   Step four: choosing the best concept.

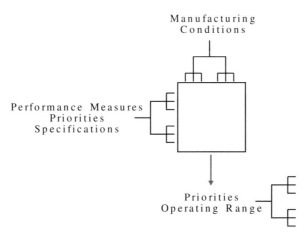

Figure 3-7   Step five: translating performance measures into manufacturing conditions.

a concurrent design process or your own unique (current) design process. For example, consider a subset of the sequential design process flow chart and the five major QFD activities. Note that you can introduce an activity associated with the QFD approach at various points in the design process (see Figure 3-8). The approach is flexible enough that your design team can decide when to start and stop the QFD process.

Adding any single QFD activity will aid in your understanding of the input you are using. You could merely apply QFD to the manufacturing process. Taking that approach, you might wind up producing a product that the customer does not want, but you will inevitably improve the manufacturing process. In general, using one of the middle or later steps of QFD will enhance your understanding, but you will gain more if you combine that activity with one or some of the earlier activities. As a rule of thumb, using any of the earlier activities is more effective than choosing one of the later activities.

DEVELOPING A DESIGN PROCESS

It is necessary and important for an organization to improve, standardize and maintain the product development process. Accordingly, an effective organization should do the following:

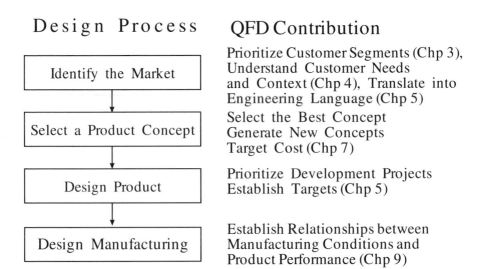

Design Process	QFD Contribution
Identify the Market	Prioritize Customer Segments (Chp 3), Understand Customer Needs and Context (Chp 4), Translate into Engineering Language (Chp 5)
Select a Product Concept	Select the Best Concept Generate New Concepts Target Cost (Chp 7)
Design Product	Prioritize Development Projects Establish Targets (Chp 5)
Design Manufacturing	Establish Relationships between Manufacturing Conditions and Product Performance (Chp 9)

Figure 3-8   QFD contributions to the design process.

- Improve the design process continuously.
- Review missing design steps.
- Know resource allocations needed for successful products.
- Review and develop cross-functional activities.
- Manage the product development process.

Creating a flow chart for all development and implementation activities is a recommended first step. Structuring the design activities is the next step in the planning process. QFD practitioners use the PDCA (Plan, Do, Check, Act) cycle (see Figure 3-9) to create a design flow. This cycle reveals the relationship between a design process flow and the organization's functions. This approach introduces a different way of looking at the design process that often generates new and useful insights.

The classic definition for the PDCA cycle is as follows:

Plan. Create a model to be tested.

Do. Try out the model.

Check. Compare the actual results obtained with the expected/predicted results.

Act. Modify or solidify (standardize/proceduralize) the theory.

Using the PDCA cycle in creating a design process requires adapting the classic definitions of these terms:

The P (Plan) integrates and translates all relevant data from the voice of the customer into design requirements (performance measures).

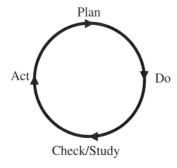

Figure 3-9 PDCA cycle.

The PDCA planning step also defines several additional activities at a finer level of detail (see Figure 3-12).

The D (Do) includes integrating technologies and concepts and confirming that the design satisfies the design requirements.

The C (Check) verifies the actual product performance.

Any differences between the team's expectations and the product's actual performance are addressed in the A (Act) step to decide whether the design cycle is completed or modifications must be made to correct the deviations noted in step C.

The PDCA cycle is a proactive process that attempts to determine what problems may exist before they occur and to create a structure that deals with the causes of those problems and prevents them from occurring. A reactive cycle analyzes problems after they have occurred. If you start the PDCA cycle with the last two steps, Checking and then Acting, you have a reactive cycle, one that resembles the traditional approach to product design. This CAPD cycle starts by checking the current performance of existing products in the particular market segment of interest.

The CAPD cycle can be used to identify the tasks needed to design a product. The process for designing a product is contained in the branches of the tree in Figure 3-10. Within the CAPD cycle, there are several steps that must be completed. Each step of CAPD has one or more tasks associated with it. For example, the Plan portion should include defining technology requirements and concept requirements. This process may also identify additional tasks besides those that have been part of the traditional design process. Though they may have been performed informally in the past, these tasks were missing from

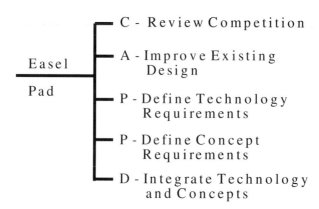

Figure 3-10 Process steps in Product Design.

the formalized design process. In creating a design process, all of the steps need to be recognized and included in a formal procedure.

RESPONSIBILITY FOR DESIGN PROCESS STEPS

Once the steps are formalized, the primary responsibility for each one must be assigned to the appropriate organizational function(s). A responsibility matrix such as shown in Figure 3-11 is useful for summarizing the design responsibilities. The shaded circle signifies primary responsibility, the open circle secondary (support) responsibility. An empty space indicates no significant responsibility. The functional group that has the primary responsibility for a step determines when it is complete and the time to move on to the next step. The functional groups that have secondary responsibility provide support to the group shouldering the primary responsibility. Job position or task description identifies who is responsible for which tasks. According to the developed design process, an organization may need to acquire new responsibilities (to address those areas not addressed in the previous design process) and to divide these new responsibilities among the functional units. The pattern of primary responsibility circles shown in Figure 3-11 also indicates that the engineering function may have too many responsibilities.

Critical Process Responsibility ○ Primary ○ Supporting	President	Marketing	Engineering	Research & Design	Quality	Manufacturing
C - Review Competition	○	●	○		○	
A - Improve Existing Design		○	●	○	○	○
P - Define Technology Requirements			●			○
P - Define Concept Requirements		○	●	○	○	○
D - Integrate Technology and Concepts		○	●		○	●

Figure 3-11 Design process steps Responsibility Matrix.

IDENTIFYING TASKS WITHIN THE DESIGN STEPS

The CAPD cycle is used to identify the key tasks for each of the steps in the design process. Each step in the CAPD cycle has one or more tasks. There are five key tasks for checking the first step in the design process (see Figure 3-12):

- *C.* Identify target market.
- *C.* Identify customer needs and expectations.
- *A.* Select performance (technical) measures.
- *P.* Design technical benchmarking.
- *D.* Evaluate competition.

Depending on the steps of the design process, the key tasks may be different. For more complex design processes, it may be necessary to use the CAPD cycle again to identify subtasks within the key tasks.

RESPONSIBILITY FOR KEY TASKS

Using the responsibility matrix for the design process (Figure 3-11), an organization identifies the functional responsibility for each key task (see Figure 3-13). The more detailed key task responsibility matrix uses the same symbols as Figure 3-11.

THE PROCESS DESIGN PROCESS CHART

The Process Design Process Chart (Figure 3-14) has rows that contain the design steps and key tasks. The columns represent the organization's functional units. Rectangles in the chart identify activities, and the location of the rectangle indicates functional participation in that activity. Arrows show the flow of documents or decisions. The column in which an output arrow is placed indicates primary functional responsibility.

The product design process chart shows the design team membership. During product development, the design team needs many documents, such as market analyses, customer requirements and specifications. These documents are normally regarded as reports within the design process. For the sake of clarity, these documents have been excerpted, and only the QFD reports are shown in the right column of Figure 3-

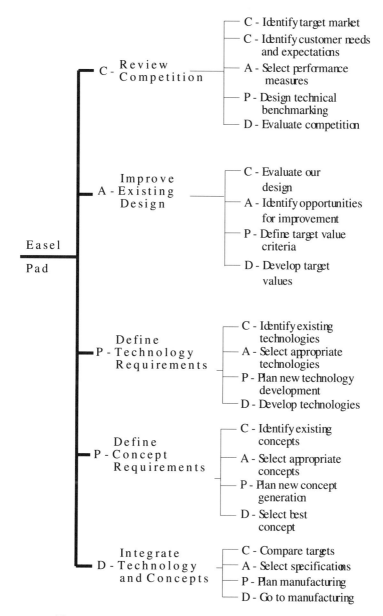

Figure 3-12 Design Process Tree (steps to tasks).

Key Task Responsibility ⬤ Primary ◯ Supporting	President	Marketing	Engineering	Research & Design	Quality	Manufacturing
C - Identify target market	⬤	⬤	◯			
C - Identify customer needs		⬤				
A - Select perform. measures		◯	⬤	◯	◯	
P - Design tech. bench market		⬤	⬤		⬤	
D - Evaluate competition		◯	⬤			
C - Evaluate our design			⬤	⬤		
A - Identify impr. oppor.		⬤	◯	◯		◯
P - Define target value criter.		⬤				◯
D - Develop target values		⬤	⬤	◯	◯	◯
C - Identify exist. tech.			⬤	⬤		
A - Select appr. tech.			◯	⬤		◯
P - Plan new tech. devel.			◯	⬤		
D - Develop technologies			⬤			
C - Identify exist. concepts		◯	⬤	⬤		
A - Select appr. concepts		◯	⬤	⬤		◯
P - Plan new concepts gen.			⬤			
D - Select best concept		◯	⬤	◯	◯	◯
C - Compare targets		⬤	⬤			
A - Select specifications			⬤			⬤
P - Plan manufacturing			◯			⬤
D - Go to manufacturing						⬤

Figure 3-13 Design process task Responsibility Matrix.

Product Design Process Chart

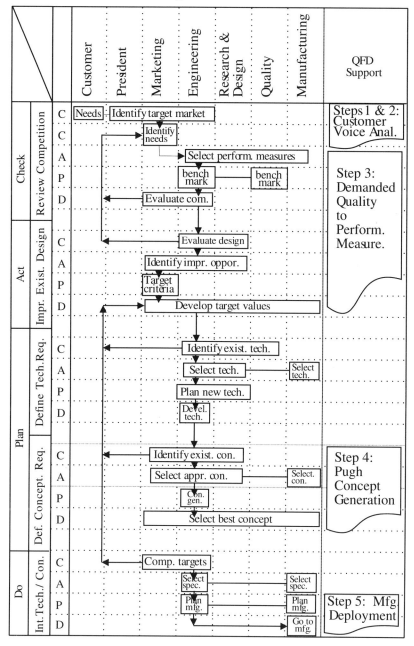

Figure 3-14   Product Design Process Chart.

14. In this example, the column contains four of the five QFD topics covered above. The Voice of the Customer analysis is done in two steps: (a) gathering the customer's demanded qualities and then (b) prioritizing those needs.

Figure 3-15 shows the overall sequence of QFD activities covered in this chapter. The product design process chart provides a detailed road map for managing a QFD project. The journey unfolds, one step at a time, beginning with the Voice of the Customer and eventually identifying the manufacturing conditions necessary to produce a world-class product.

It is possible that simply using the PDCA cycle to construct a product design process provides a significant breakthrough in thinking. This new road map for the design process may be all that the design team needs.

It is helpful to reflect on your different customer (market) segments before considering future changes in your design process. Who are your most important customers? Today? In the near future? In the distant future? How much should you listen to the demanded qualities voiced by one of your customer segments versus another? It helps to review the following key questions:

(a) What business are you in today?

(b) What business will you be in in 5 to 10 years?

Taking the design process and the QFD methodology and relating them on a step-by-step basis allow the personnel involved in design activities in an organization to add QFD methods where they feel it is appropriate. They can stop the formal QFD process whenever they have learned enough to effect a significant benefit in the design process.

Managers can show their interest and support for QFD and design activities by asking some of the following questions:

1. Which functions are represented on the design team now? What functions should be represented?

2. What customer (market) segments have been considered? How has the market been segmented? What segments are growing, which are declining? What is your market share in each?

3. What criteria have been used to rank the market segments (criteria such as ethnic mix, ease to satisfy, multiplier effect–early adopters, etc.)?

4. What breakthrough in thinking has resulted from getting close to, understanding and experiencing the customer's environment?

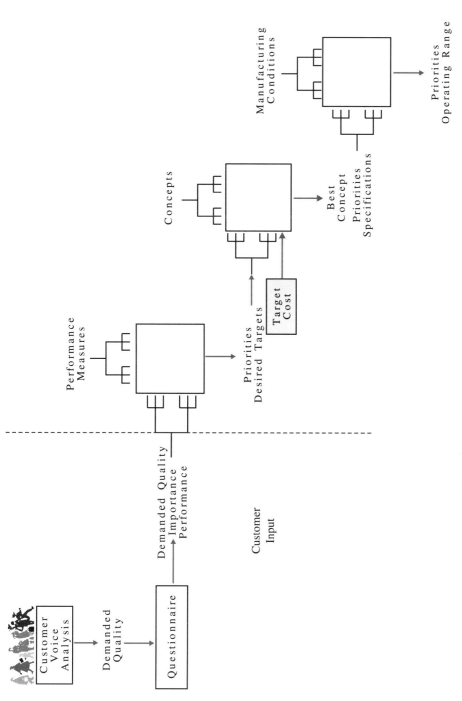

Figure 3-15 Overall sequence of QFD activities.

5. Are there any previously unknown demanded qualities?

6. Have there been any surprises in customer importance rankings and evaluations of the current product designs?

7. Are there any performance (technical) measures that are no longer needed?

8. Are there any new performance measures?

9. Are there any preexisting or new regulations that must be met?

10. Have you taken care of specific organizational requirements? Future requirements? For example, there is increasing demand for biodegradable and easily disassembled products for the German market. If the organization intends to enter or remain in the German market, today's design efforts must consider tomorrow's demands.

II

USING QFD ON A
PROJECT BASIS

4

QFD AND CREATIVITY

How is it that some products and services have the characteristics that excite customers while so many fail to even meet all the basic requirements? What distinguishes the features that will "delight" versus those that "disgust"? How can designers predict in a reliable, customer-driven fashion the features of a product or service that will position an organization ahead of its competition? Where should an organization's limited R&D resources be concentrated for maximum impact? QFD and creativity work together to answer some of these questions. By creativity we mean both innovative and breakthrough answers to the customers' needs and the customer-desired benefits.

Before we can discuss customer-driven innovation and break-throughs in products and services, however, we need to have a basis for discussing different types of quality and customer satisfaction. Noriaki Kano has developed a model of quality that is often used by QFD practitioners. The Kano Model (see Figure 4-1) describes three different types of quality. It graphs all three on a two-dimensional space divided by two axes. The vertical axis represents the degree of customer satisfaction. At the top of the vertical axis, customers are very satisfied, or "delighted." At the bottom, customers are very dissatisfied, or "disgusted." In the middle, customers are neutral, or indifferent. The horizontal axis represents the degree to which the quality is fulfilled. To the left, customer need has not been fulfilled at all, or is "absent." To the right, customer need is completed fulfilled, or fully implemented.

The first of the three types of quality is Performance Quality (PQ) and represents the "spoken," or verbalized, wants from customers.

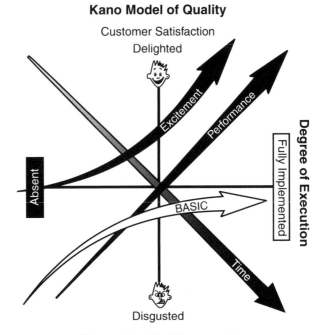

Figure 4-1   The Kano model.

These wants can be obtained through normal marketing research activities such as telephone surveys and mail-back surveys. For example, if surveyed on automobiles, a customer might ask for a fast car or an economical car, a vehicle that is comfortable on long trips or is capable of carrying a large family with all their luggage. Or, if surveyed about hotels, the same customer might indicate he or she likes to be able to check out quickly without waiting in lines and to be able to use his or her favorite long-distance phone credit card to make phone calls from the room without any extra charges. To the extent that an organization does not acknowledge, understand or deliver these PQ attributes, the customer will be very dissatisfied, or disgusted. To the degree that the organization does deliver these attributes, the customer will be very satisfied and even delighted.

One of the important aspects of PQ is that these are not features. A feature is a particular solution or way to satisfy a requirement. A "root" want or requirement is the underlying reason for the feature. The customer wants a certain benefit, not necessarily the particular feature that gives that benefit. One way to get information on what the customer expects in the way of performance requirements is by asking "Why?" to drive the feature(s) that the customer is requesting back to the root want. For example, when asked about a preference in coffee cups, the customer may request a "Styrofoam cup." An experienced researcher would recognize this request as a feature, not a true requirement. When asked "Why do they want a Styrofoam cup," the customer may say, "I want the coffee to stay hot for some time, but I don't want the cup to be too hot to hold." These are the types of performance requirements that are searched for in a QFD design project. Typical PQ characteristics will be those items commonly mentioned in advertisements or what friends would mention in a conversation. In the case of an ad for tires, it would be the price and mileage warranty. If neighbors were talking about their tires, it would be how many miles each got from their tires before they had to purchase replacements.

The second type of customer requirement is known as Basic Quality (BQ), which represents the requirements that customers will not usually talk about or even think to request ("I want a tire that is safe to drive on" or "Does your hotel bathroom have toilet paper?"). This is because from past experience they just assume that those requirements will be met. Basic Quality requirements are just expected to be there, and therefore, the customer will not express the requirement. For this reason, they are often referred to as "unspoken—unless violated." However, when violated, they become "spoken" again, usually with substantial dissatisfaction on the part of the customer as the customer is "disgusted." If BQ items are present, the customer is neutral or indifferent. They will not be delighted. For example, a customer will not be delighted after being given a *nontoxic* cup of coffee that is *black* in color (unless it was ordered with cream!) in a *nonleaking* cup. After all, all of these characteristics go without saying; they are just expected. However, if any of these are missing or forgotten, the customer will be very dissatisfied. Because BQ requirements are assumed to be there, it is difficult to get an exhaustive list of them by using elementary market research methods such as phone or mail-in surveys. To truly understand the BQ requirements, it is usually necessary to use more advanced market research methods such as focus groups and one-on-one interviews. It is also necessary to go beyond the customer and interview the personnel to whom the customer turns when things go wrong. This includes but is not limited to customer service representatives, field service personnel and regulatory agencies.

Often QFD practitioners will create Function Trees to develop the list of BQ requirements (see Figure 4-2). Technical people responsible for product and service improvements must know and have basic requirements documented and checked to ensure compliance. Often,

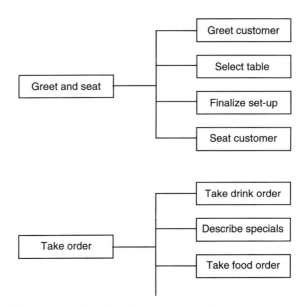

Figure 4-2 Function Tree example (for a restaurant).

many of the BQ requirements are commonsense items such as the supply of towels and soap in hotel bath rooms or that the product will not cause shock, injury or death. In the case of products such as tires, it would be that the tire would not blow out while driving or would not blow up while being inflated.

The third type of customer requirement is Excitement Quality (EQ). This type of quality is unique in that it is "unspoken" and is unexpected by customers. If EQ is absent, the customer will not be dissatisfied because it is not expected. If, however, EQ *is* present and is well implemented, it can deliver incredibly high levels of customer satisfaction. Excitement Quality can also be described as customer delights ("WOWs") or as pleasant surprises.

Some recent examples of EQ are mini-vans that have stereo and headphone jacks for the kids, cupholders that really work for a wide variety of cup sizes, restaurants that provide waiting patrons with pagers to notify them that their table is ready, televisions that help you "find" the remote control and the small Bose radio that has a remote control and sounds as if it has speakers that are much larger than the radio. Excitement Quality represents one of the keys to the development of customer-driven innovations and breakthroughs.

The Kano Model teaches us that satisfying and ultimately delighting customers requires more than just listening and responding to the customers' "spoken" current needs (PQ). One must also deliver the expected BQ requirements or be subject to customer dissatisfaction. In addition, to position the organization ahead of the competition, to gain a reputation as a leader in the organization's defined industry, everyone must be primed to anticipate and exploit any EQ opportunities that arise.

Unfortunately, the organization's leadership role due to a particular EQ feature or benefit is likely to be short-lived (unless it is linked to a broad-based patent for the product feature or the process that gives the feature). Excitement Quality, when it is implemented, is quickly recognized and copied by competitors and becomes expected by customers. It becomes just one of the regular offerings in the marketplace. Believe it or not, turn signals and heaters in cars were at one time exciting to customers (and the first Pierce-Arrow cars had only forward gears; only in the second year of production did they get a reverse!). Now these features are so basic that they are not even mentioned. For this reason, organizations must pursue a never-ending, steady process of gathering, analyzing and interpreting EQ opportunities. They need to create an environment where they constantly develop and implement new/advanced EQ into their products, services and processes over time because EQ items evolve into PQ, and PQ becomes BQ. It is essential that the organization keep the pipeline full of new and innovative benefits based on designed-in features and functions.

Some of the key methods that have been developed to identify EQ opportunities and assist in turning them into customer-driven innovation are

1. Watching customers for opportunities: Observational Research
2. Dramatic improvement in performance
3. Eliminating or improving a key product trade-off
4. Lateral benchmarking

WATCHING CUSTOMERS FOR OPPORTUNITIES: OBSERVATIONAL RESEARCH

One of the best ways to create innovative products and features is to simply watch the various types of customers as they use the product, service or process in their usual environment. This technique is referred to as Observational Research. Studying the nonverbal clues during operation and talking to the same customers are effective ways to identify opportunities for EQ.

Some clues are monitored for during Observational Research. Each offers the possibilities for product/service/process innovation:

Frustrations/Confusion

What do customers find frustrating with the product/service/process? What aspect of the product/service/process is most confusing?

Fears and Anxieties

What are customers most fearful of when using the product/service/process?

What aspect of the product/service/process causes the most anxiety?

Wasted Time, or Time-Consuming Activities

Is there any way to eliminate wasted time while using the product/service/process?

Are there any excessive time-consuming activities associated with the product/service/process?

Doing Things Wrong

Are there aspects of the product/service/process that customers misunderstand or are likely to do/use incorrectly?

Misuse, Abuse and Unexpected Usage

Can you determine situations where the product/service/process is likely to be misused?

Are customers using the product/service/process differently than its initial intended purpose?

Dangerous or Potentially Dangerous Situations
How do people get hurt using the product/service/process?

What do customers perceive as dangerous situations when using the product/service/process?

Using Aftermarket Products/Modifying Their Product

Are customers buying aftermarket products and/or modifying their product/service/process to eliminate a shortcoming or create an enhanced product/service/process?

Examining Fringe Customers

Are there fringe or niche customers that have difficulties with the product/service/process?

Are there fringe customers that could be developed into a desirable niche market?

A variety of research techniques can be used to uncover the above information:

1. Observing customers using the product/service/process in their typical environment
2. Observing and talking to customers at special events where groups of like-minded individuals gather (i.e., dog shows, auto shows, medical conventions, etc.)
3. Observing customers in a controlled environment atmosphere such as a market research product clinic
4. Observing customers in the buying/retail environment
5. Examining the product after usage and talking with the using customer
6. Talking to customers in focus groups or one-on-one interviews

A thorough understanding of customer frustrations, confusion, fears, and so on, will create a higher probability that the brainstorming for new features and solutions will truly respond to the real customer issues rather than the development of expensive high-risk, high-tech features that customers do not want. Before beginning the search for features and functions, first determine the benefits you feel the customer will value. Often it is useful to categorize the different customer sets and then brainstorm on the benefits that each would appreciate. Once the benefits have been detailed, match the features/solutions that will yield those benefits. Do not get trapped in the features game where more features are better. Features are only good where they cause the customer to derive a benefit linked to a customer want or requirement. Features that are not linked to customer benefit(s) are not value added; they are *cost added.*

A "beginner's" mindset is encouraged, which allows the wondrous ability to cut through preconceived notions and taken-for-granted assumptions to find the simple, obvious and creative. Conventional wisdom has taught us to follow rules, making it tougher to discover the exceptions that will be innovative and answer customers' needs (i.e., the breakthroughs). Observational research allows you to just watch the customer and forget the preconceived notions and traditional rules that stifle creativity.

Presented below are examples of several situations where watching the customer has led to new product-driven innovations.

Frustration and Confusion

Chrysler minivans have integrated child safety seats as an option for parents tired of dealing with the headaches associated with current child safety seats. Levi Strauss produced a line of jeans with a "skosh more room" in the rear end and thighs for baby boomer customers who had a difficult time finding jeans that fit them. Automotive manufacturers have begun to color coat the under-the-hood fluid level sticks after noticing that customers were frequently frustrated and confused when trying to locate and check their fluid levels. Sony has changed the nature of the small camcorder cassette labels with a new label that can be written on with pen or pencil and easily erased with just the swipe of a thumb. Since different people have different temperature desires when traveling on long trips, GM has developed dual-climate controls for driver and passengers.

Modifying the Product

After noticing that customers were driving cars over their brand new jeans and then dropping them into the washer and adding bleach, Levi Strauss released stone-washed, prefaded jeans. After noticing that kids were wearing ripped jeans, Levi Strauss introduced preripped jeans! Reebok noticed that many people were modifying their running shoes and tennis shoes for use in aerobics class. They introduced a line of aerobic sneakers based on the best of their customers' innovations: ankle support, extra cushioning and bright colors.

Misuse, Abuse and Unexpected Usage

Kitty litter was born when someone noticed a neighborhood cat "visiting" an open bag of clay particles manufactured to soak up oil and grease spills. After noticing that people were using orange crates and wholesale milk crates for storage containers, someone copied them in sizes suitable for everything from audio tapes to compact discs to record albums.

Wasted Time

Several restaurants (such as Outback Steakhouse) and some doctors' offices located near shopping centers provide pagers to waiting patrons.

DRASTIC IMPROVEMENTS IN PERFORMANCE

Performance Quality represents the spoken, or verbalized, wants from customers and is typical of the results generated by most market research activities. By drastically improving the level of performance quality delivered (i.e., by moving up the performance line) for a particular customer need, a product will provide high levels of satisfaction and, in some cases, even delight the customer. The best way to improve PQ is to improve your understanding of the customers' needs. There are two situations where this is useful: (a) move to the top of (or beyond) the performance quality line (resulting in customer delight), or (b) improve where everyone else is doing poorly (giving a definite competitive edge).

By examining competitive benchmarking data, you can identify not only strengths and weaknesses but also opportunities for breakthrough. When all competitors are doing poorly, many companies just write it off, saying, "We're no worse than so-and-so." In doing this they are ignoring an opportunity that, if taken, may position the company as a leader in the industry. In the case of both of situations cited above, it is possible to exploit the advantage by (a) identifying an important spoken customer requirement (focus on the root want), (b) benchmarking the competition, (c) establishing the expected performance levels, and (d) challenging the design team to dramatically improve the product substantially beyond expected performance levels.

Some examples of dramatic improvements in performance are the Marriott Hotel room check-in process, which is nearly instantaneous; concert theatres and stadiums that have two to three times as many women's restrooms as men's; GM cars using plastic door panels for superior resistance to showing door "dings" as well as dramatic increase in corrosion resistance; fiberoptic and satellite communication technology eliminating delay and static in overseas telephone conversations; and gasoline stations that allow you to save time and watch the children in the car by accepting credit cards at the pump.

The QFD process can be used together with traditional market research to gather the root customer wants. The competitive assessment diagram (see Figure 4-3) shows how the root wants can be combined with the importance rating and competitive benchmarking.

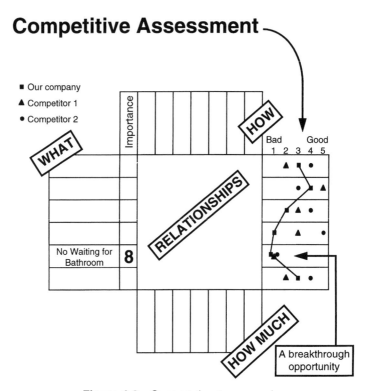

Figure 4-3   Concert theatre example.

ELIMINATING OR IMPROVING A KEY PRODUCT TRADE-OFF

A third source of innovation within QFD comes from exploring the information contained in the House of Quality. The A3 or Correlation Matrix ("roof" of the House) identifies positive and negative correlations, that is, technical trade-offs (see Figure 4-4). These trade-offs are often the source of engineering compromises because of the limitations of current technology. Although trade-offs and technical limitations are a fact of life, they represent opportunities for an R&D breakthrough. Further details on overcoming difficult situations in the roof are addressed in Chapter 9, QFD and Theory of Inventive Problem Solving.

Once a strong negative correlation is broken, it usually represents a major paradigm shift resulting in proprietary technologies and patents. After all, the best trade-off is *no* trade-off. Basically the roof of the House of Quality represents a hotbed of opportunities. QFD offers organizations the potential to drive even R&D programs with the Voice of the Customer.

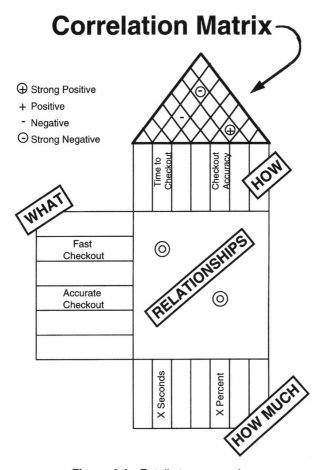

Figure 4-4 Retail store example.

Some examples of improving or eliminating trade-offs are retail stores that use bar coding and scanning technology to both increase the speed of check-out and increase their accuracy; disposable diapers that use materials that are thin and not bulky and are at the same time superabsorbent; and the latest generation of automobiles that offer both flexible performance and excellent fuel efficiency.

LATERAL BENCHMARKING

Many companies currently benchmark their competitors for a variety of reasons. In many cases, benchmarking is used as a part of the concept selection step in the QFD process. Although there are many benefits to this type of benchmarking, it usually does not result in innovation. However, if you benchmark other products and services—and really stretch yourself to learn from and explore other products and services—it can result in some creative and low-risk concepts. These concepts, if proven viable, will be perceived as innovative and exciting by customers because they represent ideas that have not been used by other competitors in your industry.

Some examples of lateral benchmarking are a large hospital chain that wanted to change the image of a hospital as being a "bad" place to go and benchmarked cruise lines to learn how they catered to their customers; Mr. Coffee coffee maker, which has the same automatically timed shut-off feature as a curling iron; and personal watercraft (such as jet skis) that should benchmark against similar recreational vehicles such as motorcycles, snowmobiles and all terrain vehicles.

In addition to the above described methods, the use of advanced market research methods with some modifications can result in innovations, breakthroughs and EQ. A method that can be effective is staging a focus group comprised of (a) "progressive," open-minded customers (sometimes called "early adopters") and (b) customer- and application-sensitive, technology-"savvy" technicians/engineers (design and manufacturing)/R&D scientists. These two groups work together in a nurturing environment with many examples of (for example) cut-aways of prototype products, drawings of failed and successful products, examples of new materials or different material combinations, working prototypes, heavy cardboard and clay mock-ups (with extra cardboard, clay, scissors, glue, etc., to construct new mock-ups or modify other mock-ups on the spot) and a computer-aided design (CAD) workstation with an experienced operator who can create and print out or update designs immediately.

Early practitioners of QFD felt that the methodology was better suited for use in designing next-generation products, and they had reservations about its applicability to breakthrough products or its use in fast-moving high-tech industries. Subsequent experience has shown that QFD not only can be very useful in those situations but also is *essential* in cases where time to market is critical. *A cautionary note:* In any design project a balance must be struck between the breakthrough and innovation aspects (Excitement Quality) and the regular Performance and Basic Quality aspects. All three of the qualities identified by Kano must be addressed simultaneously. Fortunately, there is a methodology that assists the design team in balancing the various requirements and delivering to the market a combination that reflects the customers' preferences, that is, QFD.

5

QFD AND
PRODUCT DESIGN

QFD was initially developed to assist in the design of products and is especially useful when applied to complex products. The methodology can be of great value in bridging the communications gap between the customer (who could be, e.g., the end user of a product, the person involved in its maintenance or someone linking that product to another) and the personnel involved in choosing the concept(s) as well as designing and engineering the product to be manufactured, distributed and maintained in the field.

As QFD users became more adept at applying the House of Quality, they realized that there were matrices and tables that could assist them in organizing, controlling and carrying out more of the design process with the same valuable results they had come to expect from using the House of Quality. Two approaches beyond (downstream from) the House of Quality have been popularized, the American Supplier Institute's Four Phase approach (see Figure 5-1) and the GOAL/QPC Matrix of Matrices approach (see Figure 5-2). There is also an upstream expansion, the Voice of the Customer Table (VOCT) (see Figure 5-3).

Because QFD is dynamic, flexible and expanding, there will undoubtedly be future additions to the methodology as more and more people become aware that QFD is available to assist in the design of complex products, services, processes and even systems.

There is an ongoing problem that exists when using the established approaches (ASI and GOAL/QPC): there is sufficient complexity that the design team may become confused as to which matrices need to be worked through, and in which sequence, so they can reach their goals.

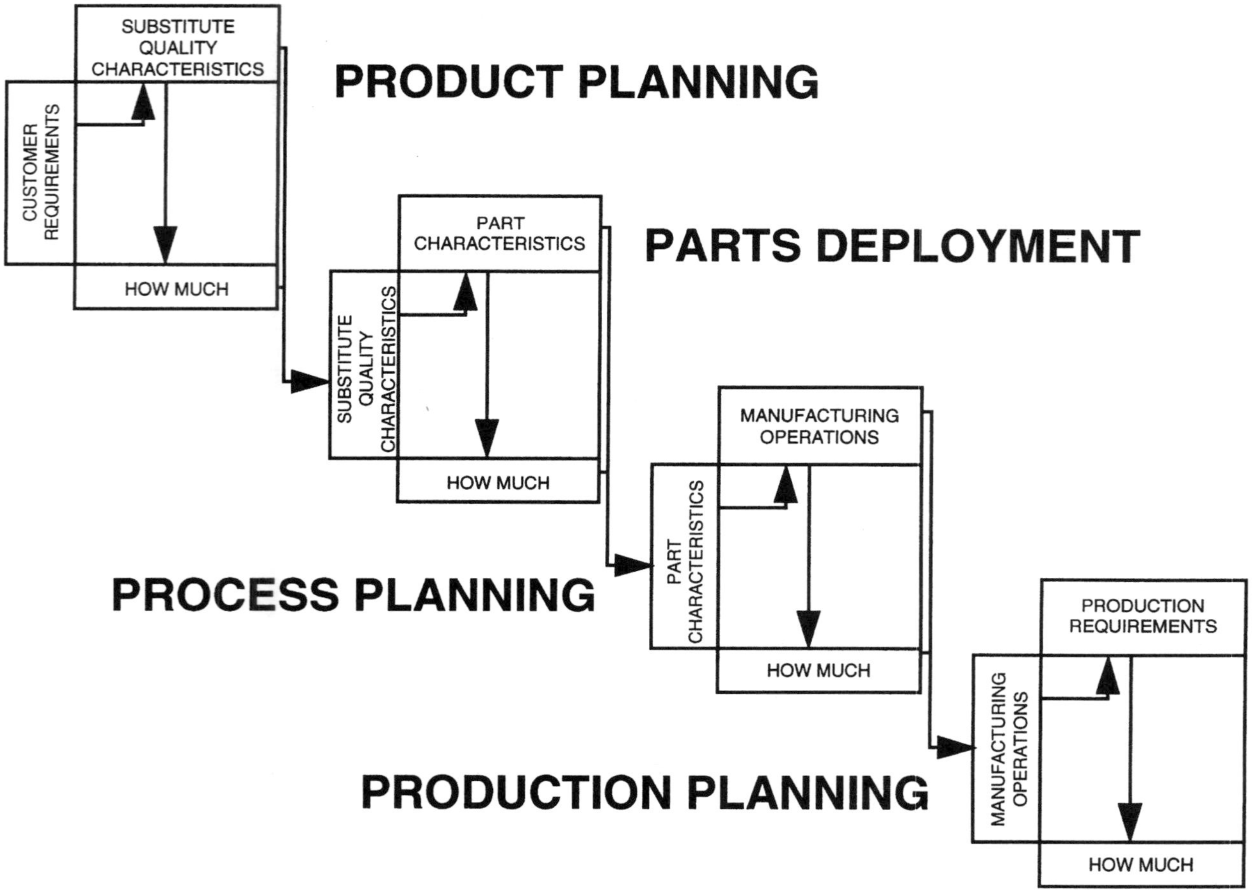

PRODUCT PLANNING

PARTS DEPLOYMENT

PROCESS PLANNING

PRODUCTION PLANNING

Figure 5-1 Four phases of QFD.

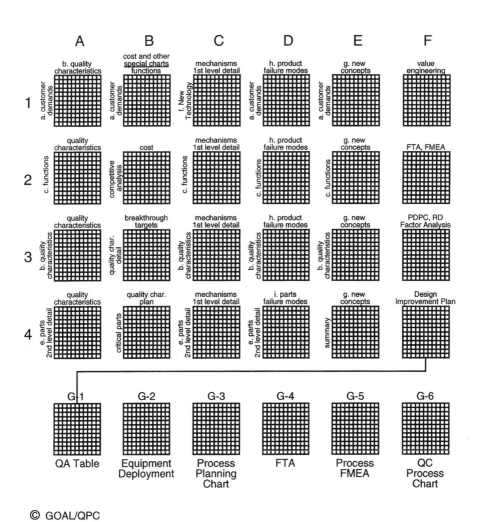

Figure 5-2   Matrix of Matrices.

Voice of the Customer	Use					
	Who	What	When	Where	Why	How
Easy to find during nighttime power failure	Adults; kids	See during power failure	Night	House; basement	See in dark; check fuses	Hold in hand; set on surface

(a)

Reworded demands	Demanded quality	Quality Characteristics	Function	Reliability	Other
Can hold easily	Can hold easily				
Can use hands free	Can use hands free				
Maintain aiming			Maintain aiming		
Fits in drawer		Diameter			
Always ready to use				Does not work	

(b)

Figure 5-3 Voice of the Customer Table (a) part 1; (b) part 2; (c) parts 1 and 2 combined; (d) part 1 portion of (c), (e) part 2 portion of (c).

To help reduce this complexity, a software program, QFD/Pathway, has been developed for design teams who want to utilize multiple matrices and who are interested in planning as much as possible so they can meet (or beat) their budget and schedule constraints. QFD/Pathway details which matrices have to be worked through to gain specific objectives and which matrices can be worked in parallel to reduce cycle time. [A copy of QFD/Pathway accompanies this book. For details on computer system requirements and installation of QFD/Pathway, see Appendix A.]

At the outset QFD/Pathway asks team members to choose which outcomes they are trying to accomplish (see Figure 5-4 for a full list of outcomes). Each desired outcome is linked to one or more matrices and the VOCT. After choosing the desired outcomes, the team initiates a search of the necessary matrices by touching the Search button. The software responds with a list of the required matrices and tables on a

Voice of the Customer Table (VOCT)

Code No.	Demographics Sex	Age	Other	Voice of Customer (Verbatims)	I/E	Who Uses It?	I/E	What Is It Used For?	I/E	When Is It Used?	I/E	Where Is It Used?	I/E	Why Is It Used?	I/E	How Is It Used?	Reworded Data	Demanded Quality	Quality Characteristic	Function	Task	Reliability (Failure Mode)	Comments
1	F	25	Single	Doesn't slip out of my hand	E	Single woman	I	To light room			E	Home			1	General	Stays in my hand	Can grip easily					Handle
																	Has a handle	Can carry easily					
2	M	25	Factory worker	Doesn't break if my kids drop it	E	Man, kids	I		1	Weekends	1	Garage	1	Fix car	1	Spotlight	Doesn't break when dropped					Breaks when dropped	
																	Bulb doesn't burn out						
3	F	35	Sales	Always works when I need it	E	Woman			1	Emergency	1	In car			1	Hand held	Works in emergencies	Works for a long time					
																		Works in emergencies					
4	F	20	Student	Turns off easily; shines a long distance	E	Student	1	Light path	1	Camping	E	Outside	i	Walking	1	Hand held, set down	Extra battery	Works anytime					Extra battery
																	Turns on/off easily	Can operate easily					
																	Lights a distance	Can turn on/off easily					
																	Lights evenly	Lights distantly					
																		Lights evenly					
																		Lights efficiently					

(I.E. = internal/external determination)

(c)

Figure 5-3 (Continued)

59

Code No.	Demographics			Voice of Customer (Verbatims)	I/E	Who Uses It?	I/E	What Is It Used For?	I/E	When Is It Used?	I/E	Where Is It Used?	I/E	Why Is It Used?	I/E	How Is It Used?
	Sex	Age	Other													
1	F	25	Single	Doesn't slip out of my hand	E	Single woman	I	To light room			E	Home			I	General
2	M	25	Factory worker	Doesn't break if my kids drop it	E	Man, kids			I	Weekends	I	Garage	I	Fix car	I	Spotlight
3	F	35	Sales	Always works when I need it	E	Woman			I	Emergency	I	In car			I	Hand held
4	F	20	Student	Turns off easily; shines a long distance	E	Student	I	Light path	I	Camping	E	Outside	I	Walking	I	Hand held, set down

Product/Service Application Context

(d)

Figure 5-3 (Continued)

(I.E. = internal/external determination)

Reworded Data	Demanded Quality	Quality Characteristic	Function	Task	Reliability (Failure Mode)	Comments
Stays in my hand	Can grip easily					
Has a handle	Can carry easily					Handle
Doesn't break when dropped					Breaks when dropped	
Bulb doesn't burn out						
Works in emergencies	Works for a long time					
	Works in emergencies					
Extra battery	Works anytime					Extra battery
Turns on/ off easily	Can operate easily					
	Can turn on/off easily					
Lights a distance	Lights distantly					
Lights evenly	Lights evenly					
	Lights efficiently					

(*e*)

Figure 5-3 *(Continued)*

Desired Outputs	Related Matrices
Better statements of demanded quality, functions, and reliability	**VOCT** [1]
Breakthrough targets and projects	**B3** + VOCT, A1
Critical control characteristics for key parts	**B4** + A1, A4
Design opportunities and challenges (functions)	**A3'** + VOCT, A1, A2, B1
Design opportunities and challenges (quality characteristics)	**A3** + VOCT, A1
Function interrelationships	**A3'** + VOCT, A1, A2, B1
Functions targeted for cost reduction	**B1** + VOCT, A1, A2
Functions with little or no customer need	**A2** + VOCT, A1
Identification of customer reliability expectations	**A1** + VOCT
Identification of unacceptable failure modes	**A1** + VOCT
Key design priorities related to customer demands	**A1** + VOCT
Key failure modes (prioritized by customer demand)	**D1** + VOCT, A1
Key parts for special controls and optimization	**A4** + VOCT, A1
Key product or service functions	**A2** + VOCT, A1
Key substitute quality characteristics related to customer demands	**A1** + VOCT
Mechanism opportunities for cost reduction	**C2** + VOCT, A1, A2, B1
Mechanisms targeted for breakthrough	**C3** + VOCT, A1, B3, C1
Missing data on key parts	**B4** + VOCT, A1, A4
Opportunities for improvement on key parts	**B4** + VOCT, A1, A4
Parts targeted for value engineering	**C4** + VOCT, A1, A2, B1, C2, A4, C1
Potential new materials to introduce	**C1**
Potential new technologies to introduce	**C1**
Substitute quality characteristic interrelationships	**A3** + VOCT, A1
Targeted manufacturing cost	**B2**
Targeted service delivery cost	**B2**

[1] Voice of the Customer Table

Figure 5-4   QFD/pathway QFD project outputs list.

single Summary Sheet. Detail Sheets can be called up by clicking on Details on the header bar. Using Next and Previous in the header bar will take the user through all of the Detail Sheets (the lower right corner indicates the number of detail sheets). The Summary Sheet also indicates which matrices can be worked on concurrently.

We now look at a hypothetical design team working on the next revision of a product they feel has some opportunity for cost reduction through value engineering. Using QFD/Pathway, the design team might select two project outputs:

Planning Summary

You selected the following QFD outputs for your project:
(Better) statements of demanded quality, functions and reliability
Parts targeted for value engineering

The selection(s) above map(s) into the following QFD matrices:
VOCT*
A1
A2
A4
B1
C1
C2
C4*

An asterisk beside any chart name indicates that you directly selected its output. No asterisk indicates a QFD chart required to provide an input to another chart.
Select DETAILS from the menu above for more information on these charts.

The following QFD charts offer an opportunity for concurrent QFD activity. This may reduce overall project cycle time. Each row introduces a combination with common inputs.

VOCT   C1
A2   A4

Select the Details option in the menu above for information on each QFD chart above.

Figure 5-5   VOCT and value engineering QFD project sequence.

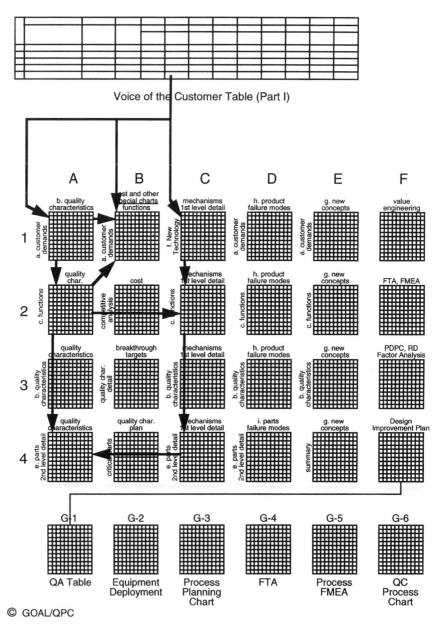

Figure 5-6   VOCT and value engineering QFD project sequence.

1. Better statements of demanded quality, functions and reliability
2. Parts targeted for value engineering

The Summary Sheet (see Figure 5-5) would indicate that the VOCT and the following matrices would be used to reach their project's desired outputs: A1, A2, A4, B1, C1, C2 and C4. The Summary Sheet would also indicate that the VOCT and C1 as well as the A2 and A4 matrices could be worked on concurrently. The resulting matrix sequence is shown in Figure 5-6.

A second design team might have a situation where they are working on a totally new product that may prove to be a breakthrough. They select three project outputs:

1. Better statements of demanded quality, functions and reliability

Planning Summary

You selected the following QFD outputs for your project:
(Better) statements of demanded quality, functions and reliability
Breakthrough targets and projects
Potential new technologies to introduce

The selection(s) above map(s) into the following QFD matrices:
VOCT*
A1
B3*
C1*

An asterisk beside any chart name indicates that you directly selected its output. No asterisk indicates a QFD chart required to provide an input to another chart.
Select DETAILS from the menu above for more information on these charts.

The following QFD charts offer an opportunity for concurrent QFD activity. This may reduce overall project cycle time. Each row introduces a combination with common inputs.

VOCT   C1

Select the Details option in the menu above for information on each QFD chart above.

Figure 5-7   VOCT and new technologies breakthrough QFD project.

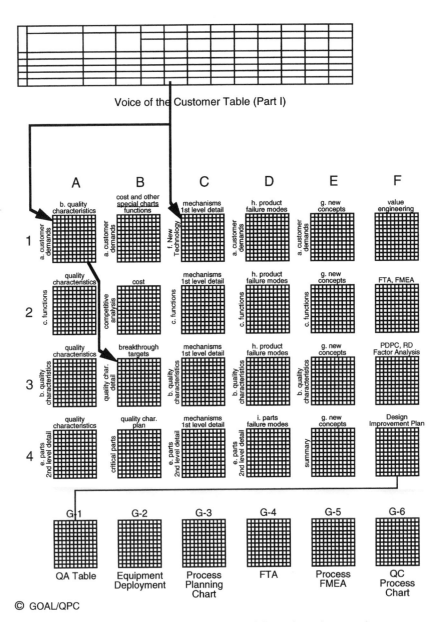

Figure 5-8   VOCT and new technologies breakthrough project matrix sequence.

2. Breakthrough targets and projects
3. Potential new technologies to introduce

The Summary Sheet (see Figure 5-7) in this case would indicate that the VOCT and the matrices A1, B3 and C1 would be used to reach the team's desired outputs, with opportunities for concurrent work in VOCT and C1. The resulting matrix sequence is shown in Figure 5-8.

6

QFD AND DESIGNING
FOR MANUFACTURABILITY
AND CUSTOMIZATION

DELIVERING THE FEATURE SET

QFD does a good job of determining the optimal feature and function set (the product definition) necessary to satisfy the Voice of the Customer. However, to be able to deliver the features and functions to the external customers (the users of the product), it is vital that the internal customers (the producers of the product) be an integral part of the product planning process. This is a strong argument for going beyond the A1 matrix and addressing not just the features and functions that will deliver the benefits the external customer wants. It is essential that how the product will be manufactured and customized be sufficiently detailed so the production processes are well defined and meet the needs of the internal customers. Only by meeting the needs of the internal customers can we be assured that the product will reach full production volume with all the features and functions intended. Let us look first at the ways manufacturability issues can impact meeting the customer's expectations and then at how these issues can be dealt with in the QFD product planning process.

In today's constantly changing environment, there are limited windows of opportunity and the information gathered from the market analyses and the Voice of the Customer activities needs to be acted upon quickly and decisively in bringing a product to market. It is important to maintain the feature/function set and *not allow features and functions to "fall off the plate" because of manufacturability problems* and the pressures of deadlines and limited resources. Despite what many managers believe and some authors reinforce, product development *does not* end at the prototype stage. Prototype technicians can lavish enough "tender loving care" on their prototypes to make them work despite deficiencies in design, tolerancing and manufacturability. Therefore, prototypes will often have the complete set of features and functions specified by the product definition.

However, if products were not designed for manufacturability, some of these carefully specified features and functions may have to be sacrificed just to get the product into production. Some companies routinely defer features and functions to the next generation when this happens. The reasons for this phenomenon are presented below in the typical order of discovery:

Cost Is Too High. If the product was not designed for low cost, then low cost is not inherent in the design and subsequent "cost reduction" efforts may recommend deleting features and functions as the only way to achieve *cost targets*. For example, Toyota decided to drop the coupe and stationwagon models to save cost for the 1997 Camry. A more insidious scenario results when part quality is compromised "to save cost," thus creating serious quality and reliability problems. This may violate customers' quality expectations and cause more features and functions to be dropped later because of low manufacturing yields. Proactive approaches to satisfy customers' cost expectation *by design* will be discussed later.

Certain Functions Will Not Work Consistently. When prototype technicians cannot get certain functions to work, they are often dropped from the feature/function set. However, sometimes functional problems do not become visible until production ramp-up if the problems have been camouflaged by the prototype technicians' tender loving care. Unfortunately, major "go/no go" decisions are often made by testing one prototype to see if *it* works without examining a statistically significant sampling size (see Chapter 7).

When the product goes into production, the statistical variations may cause so many problems that desperate measures ensue, such as compromising the feature/function set.

Tolerance Problems Increase with Multiple Sources. Going into production introduces many statistical variations from multiple sources of parts, materials and processing equipment as well as indifference of the regular workforce. More problems are encountered than with "pilot" runs that probably used single sources of supply for parts, materials and equipment. Again, the "solution" is often to compromise the product definition and leave out something.

Volume Ramps Are Too Slow. Too often, production volume ramps too slowly, because products were not designed well for manufacturability. If windows of opportunity are small, the pressure builds to get products "out the door." So, to speed the ramp-ups, production is "simplified" by eliminating features, functions or options. At the beginning of the difficult 1995 Chevrolet Lumina launch, the Oshawa plant only built white cars, "just to keep things simple." Even half a year after the launch, the plant could only offer 7 of the planned 10 colors ("Motown's Struggle to Shift on the Fly," *Business Week,* July 11, 1994, p 111.).

Yields Too Low. Low yields slow production and add quality costs. If quality was not proactively designed into products, then reactive yield improvement efforts may be necessary, which may further compromise the intended set of features and functions.

TIME-TO-MARKET

Poor product definition was the number one cause of product development delays, as reported by a survey published in the *California Management Review* (see Figure 6-1). The other major cause of time-to-market delays is solving manufacturability problems such as

- Challenges maintaining product cost, quality and reliability
- Difficulties with part fabrication, product assembly, testing, qualification, certification, and so on
- Difficulties related to the availability of unusual parts and unreliable sources of supply

An important and often neglected task of product development teams is to design products around common parts that are readily available from stable, reliable sources.

COST

Meeting customers' expectations on cost can best be achieved by *designing products for low cost* (discussed next), not by misconceptions or counterproductive measures like the following.

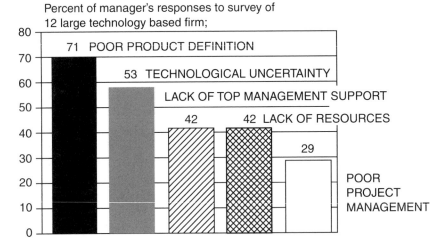

Figure 6-1 Reasons for product development delays. (*Source:* Ashok K. Gupta and David L. Wilemon, "Accelerating the Development of Technology-Based New Products," *California Management Review,* Winter 1990.)

Low-cost products do not come from

- Volume alone (this is the obsolete mass production paradigm)
- Cheap parts
- Cutting corners; omitting features and functions; compromising the product definition
- Cost reduction after design (80–90% of cost is committed by design and cannot be easily removed; shaving cost off the *reported* cost might explode other costs)
- Later redesign efforts to lower cost (early "cast-in-concrete" decisions limit opportunities, redesign efforts cost money and time and the redesign probably will never happen)
- "Saving" cost by cutting product development and continuous improvement efforts
- Manufacturing misconceptions such as purchasing policies based on low bidder and using off-shore manufacturing to lower labor costs

Pursuing any of these strategies can cause significant counterproductive effects. Specifying cheap parts and low bidders may appear to minimize *reported* cost but can easily explode the total costs of quality and reliability. "Cost reduction" efforts after the design is completed may not even pay for themselves over the life of the product and yet waste valuable resources that could have been invested in designing products to be low cost by design. Departmental cost cutting *across the board* will compromise product development and continuous improvement efforts and, thus, increase other real and opportunity costs.

Off-shore manufacturing is often touted as a way to "cut labor costs in half." However, labor is becoming a smaller and smaller portion of product cost especially with products designed to minimize assembly cost. Labor productivity differences often cancel out any proposed labor savings. Long "pipelines" to markets or subsequent operations cause delivery delays and significant quality costs because of long and slow feedback loops. Off-shore manufacturing also incurs a plethora of other costs for start-up, travel, communications, shipping, transfer costs and costs related to recruiting, training and staffing. When off-shore plants are added instead of replacing existing operations (which is often the case to preserve the original plant as a pilot plant), there will be extra costs for duplication of infrastructure (for personnel, payroll, training, etc.) and equipment, since all plants need the minimum "set" of equipment to build any products, whether they are pilot or production plants.

Ironically, automation and process improvements are hard to justify whenever there is such a *low labor cost.* Thus, off-shore operations often miss out on the quality and throughput advantages of automation and other process improvements. Finally, separating engineering from manufacturing devastates *concurrent engineering,* which is really the ultimate source of low-cost products in the first place.

MEETING CUSTOMERS' COST EXPECTATIONS

Designing to Minimize Material and Labor Cost

Low material and labor cost is determined by such low-cost design techniques as Design for Manufacturability (DFM) (Anderson, 1990), Robust Design using Taguchi Methods for optimizing tolerances to achieve high quality at low cost (see Chapter 7) and concept/architecture optimization. Studies have shown that 60% of a product's lifetime cumulative costs are committed by the end of the concept/architecture phase, as shown by Figure 6-2 (Anderson, 1990, p. 16). By the time the design is completed, 80% of the cost is committed.

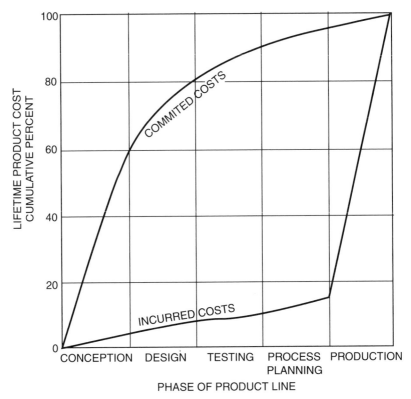

Figure 6-2 Product costs committed by phase.

Design for Manufacturability (DFM) Overview*

In the Four Phase approach to QFD (see Chapter 23, Alternative Approaches to QFD), the first phase converts customer requirements to design requirements (the Product Planning Phase), and the second phase converts design requirements to part characteristics (the Parts Deployment Phase). It is in the second phase where design requirements such as low cost (both for the parts and for the assembly operations), reliability and consistency come into play. This Parts Deployment Matrix has the inputs of design characteristics (the WHATs) and outputs of part characteristics (the HOWs). What follows is a listing of ideas (the HOWs) on how design teams can address the WHATs of cost, reliability and consistency in the Parts Deployment Matrix.

---

* This section is from a seminar, "Design for Manufacturability," by D. M. Anderson, Anderson Seminars and Consulting.

Avoiding Left/Right-Hand Parts. Avoid designing mirror image (right- or left-hand) parts. Design the product so the same part can function in both the right- and left-hand modes. If identical parts cannot perform both functions, add features to both right- and left-hand parts to make them the same.

Another way of saying this is to use "paired" parts instead of right- and left-hand parts. Purchasing of paired parts (plus all the internal material supply functions) is for *twice* the quantity and *half* the number of types of parts. This can have a significant impact with many paired parts in high volume. At one time or another, everyone has opened a briefcase or suitcase upside down because the top *looks* like the bottom. The reason for this is that top and bottom are *identical* parts used in pairs.

This principle saves cost and time in three important ways, compared to right/left situations:

- Part variety is cut in half, which is especially important for flexible operations.
- Purchasing quantities are doubled, resulting in purchasing leverage.
- Tooling costs are cut in half, which can be a very important consideration for complex permanent dies and molds for casting and plastic molding.

Designing Parts for Symmetry. Design each part to be symmetrical from every "view" (in a drafting sense) so that the part does not have to be oriented for assembly. In manual assembly, symmetrical parts cannot be installed backward, a major potential quality problem associated with manual assembly. In automatic assembly, symmetrical parts do not require special sensors or mechanisms to orient them correctly. The extra cost of making the part symmetrical (the extra holes or whatever other feature is necessary) will probably be saved many times over by not having to develop complex orienting mechanisms and by avoiding quality problems: It is a little known fact that in felt-tipped pens the felt is pointed on *both ends* so that automatic assembly machines do not have to orient the felt.

Making Parts Very Asymmetrical If Part Symmetry Is Not Possible.
The best part for assembly is one that is symmetrical in all views. The *worst* part is one that is *slightly* asymmetrical and may be installed wrong because the worker or robot could not detect the asymmetry. Or worse, the part may be *forced* in the wrong orientation by a worker

(that thinks the tolerance is wrong) or by a robot (that does not know any better).

So, if symmetry cannot be achieved, make the parts very asymmetrical. Then workers will be less likely to install the part backward because *it will not fit backward.* Automation machinery may be able to orient the part with less expensive sensors and intelligence. *In fact, very asymmetrical parts may even be able to be oriented by simple stationary guides over conveyor belts.*

Using Poka-Yoke to Mistake Proof Designs. "Poka-yoke" is Japanese for mistake proofing, a concept applied mostly to preventing mistakes in factory operations. Designing symmetrical parts is a poka-yoke principle, since symmetry prevents incorrect assembly. Part standardization (the next point) has a poka-yoke effect since having fewer part types results in less probability of using the wrong part. Parts that must be different should be *very* different. For instance, different gears can be assembled to different-diameter shafts. Shafts that need to be different should have different diameters or lengths to avoid assembly mistakes.

Note that once poka-yoke is designed into the product, it ensures quality for the life of the product, unlike inspections and process controls, which must be on-going efforts.

Standardizing Part, Features, Tools, Raw Materials, Tooling, Fixturing Geometries and Processes. Select parts from the company preferred parts lists. Reuse parts that are prevalent and have been proven to be reliable in previous designs. Use off-the-shelf hardware produced in volume by several vendors. Standardize on design features like drilled hole sizes, punched hole shapes, thread types, bend radii, etc. Each different size requires the factory to have a unique tool in stock and to set it up for each usage.

Using Off-the-Shelf Parts. Anderson's Law states: 'Never design a part you can buy from a catalog.' Rarely can anyone save any money or time by designing and building parts that are available off the shelf. Internally produced parts usually cost much more than company accounting systems indicate. Internally produced parts usually take more time to get into production.

Eliminating Adjustments/Calibrations. Products should not need any mechanical or electrical adjustments or calibrations, unless required for customer use. Products should be designed so that there are no adjustments required in assembly. Adjustments slow down the assembly process and can cause quality problems if not performed correctly. *Zero adjustments* should be a goal for the product design; the design team should use all the creativity at its disposal to achieve that goal. If adjustments are required for customer use, there should be a "default" setting that is easy to set during manufacture, for instance, at a detent or a setting at one of the extremes of the adjustment. Calibration is a form of adjustment that is time consuming and may require expensive equipment. Often calibrations can only take place after the product is fully assembled, making correction difficult.

Providing Access. Provide unobstructed access for parts *and* tools. Each part must not only be designed to fit in its destination location but also have an assembly path for entry into the product. This motion must not risk damage to the part or product and, of course, must not endanger workers.

Equally important is access for *tools and the tool operator,* whether this is a worker or a robot arm, which usually requires more access room than a worker. Access may be needed for screwdrivers, wrenches, welding torches, electronic probes, etc. Remember that workers may be assembling these products all day long and having to go through awkward contortions to assemble each product can lead to worker fatigue, slow throughput, poor product quality and even worker injury. Access is also needed for field repair, where the tools may be simpler and perhaps thicker.

Making Parts Independently Replaceable. Products with independently replaceable parts are easier to repair because the parts can be replaced without having to remove other parts first. The order of assembly is more flexible since parts can be added in any order. This could be a valuable asset in times of shortages, in which case the rest of the product could be built and the hard-to-get part added when it arrives. If this is not possible, order assembly so that the most reliable part goes in first and the most likely to fail goes in last. If parts must be added sequentially, make sure that the most likely to fail are the easiest to remove. This is important for both factory assembly *and* field repair because parts can fail in factory test or in the field.

Adhering to Specific Process Design Guidelines. This is especially important for specific processes such as welding, casting, forging, extruding, forming, stamping, turning, milling, grinding and plastic molding. Some reference books are available that provide a summary of design guidelines for many specific processes (Bralla, 1986; SME, 1992; Trucks, 1987). Many specialized books devoted to single processes are also available.

Minimizing Tolerance Demands by Design. For instance, a shaft should be retained longitudinally by a single alignment feature, not one at each end, since this would induce unnecessary tolerance demands. Locating parts by two round pins in two round holes makes the tolerance between both holes *and* both pins critical. The DFM approach would be to locate the part by two round holes that fit on one round pin and one "diamond" pin, as used for machine shop tooling fixtures.

Combining Parts. Parts should be combined for ease of assembly and tolerance control. One way to reduce the number of parts is to combine parts and functions into a single part. Purchased parts can be ordered combined; for instance, washers and lockwashers can be combined with bolts, screws and nuts. Plastic or cast parts can have labels and serial numbers molded or cast in. Designers should strive for multifunctional parts rather than multiple parts. Parts should be combined whenever possible. The criteria for combining parts are presented as follows:

- When the product is in operation, do adjacent parts move with respect to each other?
- Must adjacent parts be made of different materials?
- Must adjacent parts be able to separate for assembly or service?

If all three answers are no, consider combining the parts into one. It is important to remember that every interface between parts requires geometrical features to be manufactured and their tolerance to be held on both sides. *Eliminating interfaces eliminates the need to create interface features and to hold their tolerances.*

Designing to Lower Overhead Cost

Product development costs can be minimized with good product development practices, as presented in what follows.

Optimal Selection of Products to Develop. This practice focuses on product families that share parts, modules, software, processes, engineering and logistics support. Decisions should be made in terms of product families, or platforms, and their evolution *over time.* An important initial criteria is the *true* profitability of existing products.

Multifunctional Design Teams with All Specializations Present and Active Early. This is the key to success of concurrent engineering, that is, making sure that all the specializations are available and making useful contributions to the team effort *early in the design cycle.* Far too often, this is the plan, but in practice too many people are so busy "fighting fires" on the last product or other daily emergencies that they are not able to design products to eliminate the need for fire fighting in the first place.

Methodical Product Definition. The intent of this practice is to satisfy the Voice of the Customer *by the first design.* This avoids the expense and delays associated with a change in the design midstream because "customers changed their minds." This really means that the initial product definition failed to capture the true Voice of the Customer.

Raising and Resolving Issues Early. It is much easier to raise and resolve issues early than after much of the design has been cast in concrete. Issues are not limited to functional or processing feasibility. They include issues related to manufacturability, serviceability, certification, qualification and regulatory compliance. Issue analysis includes risk analysis to determine the probability and consequences of what might go wrong. The formal process for this is called Failure Mode and Effects Analysis (FMEA), which is the subject of Chapter 8. [For discussions on how Japanese companies raise and resolve issues early, see Womack et al. (1990, Chapter 5, p. 115) and Dertouzos et al. (Chapter 5, p. 70).]

Simplifying Concepts and Optimizing Product Family Architecture. As noted earlier, the concept/architecture phase determines 60% of a product's lifetime cost. The key to smooth and easy launch into production is thorough up-front work to methodically define the product, raise and resolve issues, simplify concepts and optimize product architecture.

Avoiding "Reinventing the Wheel." This relates to every product development with maximum use of reusable engineering and versatile modules. It also minimizes debugging costs by using existing modules that have already been debugged. One of the appeals of object-oriented software is the availability of debugged modules for reuse.

Basing Designs on Off-the-Shelf Parts and Eliminating Cost of Designing Those Parts. There is an incredible array of off-the-shelf parts available. However, product architecture may have to be predicated on early decisions to maximize off-the-shelf hardware. For instance, if the architecture of a complex electronic system can be based on standard 19-inch rack system cabinetry, then the design team does not have to design any of the following components: the cabinet, doors, hinges, latches, partitions, card cages, drawer slides, fans, filters, etc. If standard busses and card cages, such as VMEbus and Multibus II, are specified by the product architecture, then standard printed circuit boards can be purchased off the shelf for processing, input/output, memory, networking, and so on.

Basing Decisions on Rational Criteria. This practice is concerned with methodical product definition, total cost accounting (see discussion below), the real time-to-market (to stabilized production) and design of experiments (see Chapter 7). Arbitrary decisions must be avoided, such as arbitrary fastener selection, tolerances, overhead allocation algorithms, project milestone deadlines and detailed design decisions. These decisions are often arbitrary if engineers fail to design the product as a system or ignore any design considerations, such as cost, manufacturability, testability, quality, reliability, serviceability, repairability, human factors, safety, aesthetics, ergonomics, environmental issues, regulatory compliance or means to customize and upgrade products.

"Do the Right Things Right the First Time." This practice focuses on following the "map" provided by the QFD matrices to minimize change orders, problem solving, fire fighting and redesign, the ultimate waste of engineering resources.

Extended Product Lives through Upgrades Rather Than Redesigns. One of the benefits of mass customization (discussed later) is the ability to do *ultrafast* time-to-market, where new products are really planned *variations on a theme* based on common parts and modular product architecture. This is much faster than is otherwise possible with independent products that do not benefit from product–family synergies in design and manufacture.

Avoiding Diagnostic Test Development. This is accomplished by designing quality into the product and then building it in with process controls. IBM has calculated that if its printed circuit board operations can achieve a First-Pass-Accept (FPA) rate above 98.5%, then it can dispense with diagnostic test development, which for some products can exceed the cost and calendar time of product development. Further, eliminating diagnostic testing also eliminates the need to utilize sophisticated Automatic Test Equipment (ATE), which often cost over $1.5 million. At the break-even threshold (e.g., IBM's 98.5%) the cost saved in test development and test equipment exceeds the cost of throwing away the 1.5% of the product that fail the final functional test.

Minimizing Vulnerability to Market Shifts. Shorter product development cycles result in less chance of market shifts and technical obsolescence by the time the product reaches the market.

Minimizing Manufacturing Overhead Cost. This means designing to minimize overhead costs:

- Designing for Just-in-Time can virtually eliminate set-up, kitting and Work-in-Process (WIP) inventory costs and minimize floor space, material overhead costs, Bill of Materials (BOM) and inventory-related costs.
- Designing for Build-to-Order can eliminate finished-goods inventory and many distribution costs.
- Designing for Flexible Manufacturing can maximize machine tool utilization and factory throughput.
- Standardization of parts and raw materials can minimize material overhead and simplify spare-parts logistics.
- Designing for easy customization can minimize customization and configuration costs (see next discussion on mass customization).

Minimizing Cost of Quality by Design. Advanced product development can *design in quality*. Concurrently engineered production processes that are in control can *build in quality* (see Chapter 7).

Total Cost Measurement

Usual Definition of Cost. Traditional cost systems provide the cost breakdown (shown on the left in Figure 6-3) and encourage product development teams to focus only on material, labor and tooling costs. Considering only these costs gives a limited perspective and can lead to short-sighted conclusions that 80% of the product's cost is parts

Figure 6-3 Cost breakdowns: typical compared to total cost.

(and tooling). Therefore, cost reduction measures are often focused on minimizing parts costs, usually by buying cheaper parts.

Total Definition of Cost. Total cost measurements provide the cost breakdown (shown on the right side of Figure 6-3) and make it possible for decisions to be based on total cost considerations. [See Hicks (1992) for a quick and cost-effective way to implement total cost measurement.] Designers must acknowledge overhead costs are significant (often more than labor and materials) and that they have influence over overhead costs. Some engineers think manufacturability is not important if they think that labor cost is a small percentage of the total cost. However, manufacturability problems can cause significant overhead demands for problem solving, fire fighting, Engineering Change Orders (ECOs), documentation support, tooling changes and penalties and missed opportunities caused by delays. [See Anderson and Pine (1997, Chapter 6) for a thorough discussion of how total cost measurement can help product development teams design low-cost products.]

MASS CUSTOMIZATION

In defining the external customer there is often difficulty because there are, in fact, not one but several customer groups. The product design team must then grapple with the decisions inherent with diverse customers. Just how much of the team's efforts should be dedicated to this customer group versus that customer group? Should the product be designed with enough capability to handle the worst case? Even if that means extra expense for all of the other groups? Should the design team compromise on the product features and run the risk of alienating one or more of the customer groups? Is there some other way to resolve the conflicts between the voices of several different customer groups, all of whom are going to use the product but in slightly different modes?

Fortunately, there is a way to deal with this problem: give each customer group their own version of the product. Even better, give each customer a uniquely customized product. Most manufacturers are forced to customize products to some degree for increasingly selective customers. These customers are buying in a marketplace with an increasingly large selection of products (and variation within product groups). In other situations, a manufacturer may want to adapt a standard product to compete in a niche market. However, if products are not designed well to allow for this, and if manufacturing is not flexible enough, then customizing products can be a slow and costly ordeal.

When thinking of customization, the design team needs to look closely at the A1 matrix's inputs (the WHATs) and determine how many priority reversals there are (between different customer groups), by how much and how many variations in targets there are and the associated costs to address the entire spectrum of targets. If there are competitive reasons to design a spectrum of different but related products, then the Voice of the Customer and the A1 matrix need to be split to address a spectrum of separate products. For instance, at one end of the spectrum, there might be economy, low-end products concerned almost exclusively with initial cost, while the other end of the spectrum might be commercial high-usage products concerned almost exclusively with operating cost and reliability. The design team can serve all these markets by designing one versatile product *family* around modular product architecture, common parts and flexible customization scenarios.

The next paradigm, following the century-old Mass Production, is *Mass Customization* (Pine, 1993), which is the ability to design and manufacture customized products *at mass production efficiency and speed.* This section will discuss how to develop products that can be quickly and inexpensively customized for niche markets or even for individual customers.

Success will depend on strong product development methodologies that enable multifunctional design teams to concurrently design whole families of products and Flexible Manufacturing processes. Products must be designed around common parts, versatile modules, standardized interfaces, common fixturing geometries and standard processes (Anderson and Pine, 1997, Chapter 5). To assure the required manufacturing flexibility, products and processes must be concurrently designed to eliminate all set-up steps, such as kitting, retrieving parts or tools, changing fixtures, manually positioning parts and finding instructions (Anderson and Pine, 1997, Chapter 7).

The results will be well worth the investment. Companies that use these technologies will experience ultrafast development of modular products, the agility to quickly respond to changing markets and opportunities, early market penetration with the associated higher sales, delivery superiority, premium pricing opportunities, lower overhead costs and better satisfaction of customers' needs.

Product Definition for Mass Customization: The Voice of Many Customers

For Mass Customization, all products in the mass customization family must be defined so that customers can actually receive custom products. Thus, the product definition is not a single definition but a *range* of product definitions that represent various combinations of modules, standard parts, custom parts, custom configurations and customized dimensions.

To design products and processes for Mass Customization, the *scope* of the product family breadth must be understood. It is important to establish the *optimal* scope of product family customization. It may not be necessary or even feasible to offer every possible variation of products in a product family. Since customers may only need certain variations (as indicated by the ranking of their inputs on the A1 matrix), it would be a waste of effort to offer more variety than is really appreciated. Thus, mass customization product families must be defined with the optimal scope of customization.

QFD for Mass Customization

Mass Customization needs a methodology that will translate the voice of the customer for families of customized products (Anderson and Pine, 1997, Chapter 10). Figure 6-4 shows the QFD chart for Mass Customization with the "design specs" expanded into five rows. The first new row, "to be customized by," indicates if the product is to be customized in the factory (F), at the dealer or distributor (D), at the point of sale by the user (P), by the user's technical staff (T) or by the actual user (U) or is self-adjusting (S). Other symbols representing other customization scenarios can be used as appropriate. A blank or a question mark could mean "to be determined."

The purpose of these entries is to identify where and by whom the customization will be done. The design team will have to design the

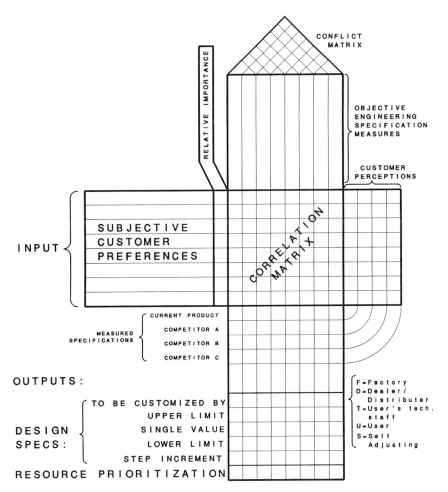

Figure 6-4 QFD for Mass Customization

product for compatibility with this customization strategy. Factory customization would rely on flexible manufacturing to be able to quickly build any configuration that is called for. Dealer/distributor customization would have to take into account the tools, parts, patience and skill levels present in *all* dealers and distributors. User customization would have to be designed for the anticipated range of user' skill levels, patience and tools needed, if any. A variation of user customization would be a situation common in large companies where technical staffs or departments would perform the customization, for example, configuring computers for all office workers or customizing production tools for factories. Self-customizing products literally adjust automatically to "fit" the user.

The next three rows contain the "single-value" row, which would be used for those engineering specifications that can be expressed by a single value and not a range of values. The upper and lower limits specify the range of dimensions for applicable specifications. The step increment would specify the steps or increments that could be appropriate for values within the range. A zero entry would indicate infinitely variable dimensions between upper and lower limits.

As with discrete QFD, the design team would base the design on the design specification target values or ranges and the steps. Understanding how each range would be customized will help develop the resource prioritization.

Benefits of Mass Customization

Better Satisfied Customer Needs.   This is possible with customized products that are exactly what customers want or at least are more closely defined than any competitor. The increase in sales would be the difference between customers that would be satisfied with standard products and customers that would be satisfied with custom products.

Price Premiums for Customized Products.   This occurs because customized products better meet customers' needs and, therefore, should be worth more to customers. The customized product, in the form the customer wants, may have little or no competition and, thus, may enjoy somewhat of a monopoly with all the ensuing pricing advantages (at least for a while).

Broader Markets by Versatile Designs and Flexible Plants.   Even if products are not customized for individual customers, flexible operations can efficiently produce products for many small niche markets that might otherwise be ignored. Common modules can still enjoy economies of scale even if the products compete in many low-volume markets. Companies that are organized to pursue Mass Customization will be able to compete better in niche markets than less flexible competitors.

Saving Customers Money By Not Needing to Buy Unwanted Options. Because customized products can be ordered with only the options customers want, they will not be forced to buy a "bundled" option package to obtain the one option they really want. Even with a premium price, customers may still save money by avoiding unwanted options. This is a real win–win situation if manufacturers can charge a premium and customers still pay less for exactly what they want. In addition, flexible, modular designs give customers better choices regarding options that may be added later as separate purchases.

Good Market Research. Customizing some products can predict customer preferences for related mass-produced products. Standard products will be quicker to market because customized products will be anticipating markets better. For example, National Bicycle's mass-customized bicycle operations (Moffat, 1990) offers an almost infinite variety of colors to its customers. The mass customization operation records the most popular colors ordered and feeds this information next door to the larger mass production operation which offers those colors on standard models.

7

QFD AND ROBUST DESIGN

Genichi Taguchi has developed a collection of pragmatic methods that lead to the discovery of how to reduce product and process variability. These methods represent a system that can be applied during R&D, product design, and manufacturing. As a result, products can be rapidly developed that perform well at high-quality levels and low costs. Taguchi calls this system Quality Engineering, although most people outside of Japan refer to it as Taguchi Methods.

This approach takes a different view of three major issues regarding product quality:

1. How to evaluate quality
2. How to efficiently improve quality and cost
3. How to cost-effectively monitor and maintain quality

HOW TO EVALUATE QUALITY

A company is in business to make a profit. Its products must be competitive with those of its competitors to maintain the financial health of the company. The relationship between profit, sales and cost can be represented by

$$Profit = sales - cost$$

Profits are increased by increasing sales and/or decreasing costs. Quality improvement is the most effective way of achieving this simultaneous increase in sales and reduction in costs. This may seem paradoxical. Our first reaction is that an increase in quality means an automatic increase in costs.

Cost and quality are linked in the same way customers evaluate products. The way quality is evaluated must take this relationship into account. It must allow for continuous improvement in quality and reduction in cost. A company improves quality to remain competitive. We need to be able to evaluate quality in a way that reflects corporate objectives. Most Western companies use the following measures as indicators of product quality: Number of Defects, Percent Defective, Number of Failures, Mean Time between Failure, Scrap/Rework Cost, Percent Yield, Process Capability Indices, Warranty Returns or Costs, Field Service Information, Competitive Analyses, and so on.

However, these indicators detect poor quality after the fact. These measures are cost related and tell us that action is needed. Therefore, they are important for management to monitor. However, from a technical standpoint, they are poor indices to use for assessment of activities to develop or improve quality. In reality these measures are only the symptoms or effects of poor product design or manufacturing process functions or they are the results of variability of the intended product or process function. For example, it is not the function of an automobile's brake to squeal. It is not efficient or effective for engineers to measure the squeal to reduce the squeal.

A useful evaluator of quality that can help engineers in their quality improvement activities must satisfy the following criteria:

(a) It can be used in upstream quality improvement. Many of the usual measures of quality rely on postdesign or postproduction figures. These are available only after the product has been designed and tested or manufactured. By that time it is too late to improve quality. We must be able to evaluate quality in upstream stages, that is, during R&D as well as product and process design.

(b) It must appropriately express customer and engineering objectives. Our measure of quality must truly express the quality of the product/process in terms of engineering intent as well as customer response to product function. It must be able to evaluate quality in terms of ideal product function.

(c) It must express the relationship between cost and quality. It must include the effects of costs that arise because products fail to function as expected in the hands of the customer. These costs affect market share and long-term profit.

Taguchi has proposed two upstream evaluators of quality that effectively meet these criteria: (1) the Quality Loss Function (QLF) and (2) the Signal-to-Noise Ratio (S/N). Every product, process or service has a specific function and an ideal performance target. This represents the engineering intent of the product to which customers respond. Quality Engineering efforts focus on reducing variability around the ideal. The smaller the variability, the higher the quality. Both the QLF and the S/N are measures of this variability.

The QLF converts this variability into monetary terms that represent an estimate of the costs a customer incurs when a product fails to function as intended. These costs can include scrap/rework costs, the cost of replacing a failed product, the time and money spent for repair, etc. These are factors that affect market share. Taguchi calls these costs a loss (i.e., a loss to society). The QLF provides a quantitative monetary estimate of this loss.

The QLF thus evaluates quality in terms of the ideal product function and customer response. It can be used in the earliest stages of product planning and development as well as in production. Because it evaluates quality in monetary terms, it can be used effectively to solve cost–quality trade-off problems. This is done by finding the balance between cost and quality that can increase profit. Improving quality cost effectively reduces Loss. In management terms: the objective of quality improvement is to reduce the loss.

The S/N also measures variability around the target performance. It is a measure of the stability and reliability of performance in the face of those uncontrollable factors (i.e., causes of variation) that a product will encounter on the factory floor and when used by the customer. The S/N gives the design engineer a reliable upstream estimate of how a product will function at downstream stages. It evaluates quality in terms of decibels (dB) and serves as an ideal characteristic for measuring quality improvement in engineering terms. In engineering terms, the objective of quality improvements is to increase the S/N.

Whatever the system is, whether material, a product or manufacturing process, an "intent" can be identified. The intent is what the system is designed to do. The intent of an automotive brake is to smoothly slow down a vehicle. The intent of an injection molding machine is to create an intended shape (and dimensions) with a certain property. To achieve an intent, any engineered system uses some form of energy transformation. Assuming the intent is what customers want, any quality problem is a symptom of variability in this energy transformation. If we are interested in developing a high-quality product, it is far more efficient and effective to reduce the variability of the energy transformation by maximizing the S/N than it is to reduce problems by measuring the symptoms .

A typical design process is to repeat the activities of design–build–test. First, the product/process concept is developed and a prototype is built. Then, the prototype undergoes testing to discover problems by observing the symptoms (i.e., failures), it is redesigned in an attempt to reduce/eliminate failures and it is tested again.

Redesign–build–test–redesign–build–test will be repeated until either the deadline comes for release of the product or the design is deemed good enough. In this approach, we usually are not evaluating the variability of the intended function. What we are measuring are symptoms. Then what typically happens is that solving one type of problem, or failure mode, introduces another.

Contrast the above problem-solving approach with problem solving using the S/N. Conceptually, the Signal-to-Noise Ratio is represented by

$$S/N = \frac{\text{energy transformed to perform the intended function}}{\text{energy transformed to other than the intended function}}$$

This is illustrated in Figure 7-1 using a brake system as an example. Squealing is not an intended effect of a quality brake system. One may be able to reduce the squealing by measuring the noise (this is not easy due to the strong interactions that reside among the various factors), but it may not necessarily improve the S/N. In fact, it may worsen other symptoms (or failure modes) or it may even introduce new symptoms. What should be measured to reduce the squealing?

The theme of the 1989 Taguchi Methods Symposium sponsored by the American Supplier Institute was "If you want quality, don't measure quality." This way of thinking is very important for the productivity of engineering activities. In the case of a brake system, its function is to create torque, which causes deceleration. We should measure the relationship between the force of the driver's foot on the brake pedal (the signal) and the torque generated (the output response). Then, we should evaluate the variability of this relationship using the S/N. This relationship and the concept behind the calculation of the S/N are shown in Figure 7-2.

Variability causes various symptoms such as squeal and vibration. Because S/Ns are logarithms, increasing the S/N by 6 dB is equivalent

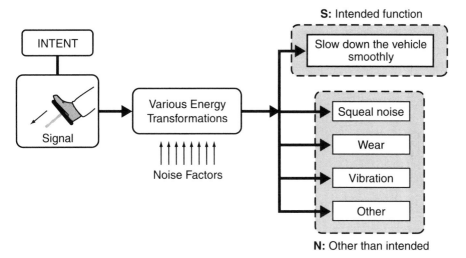

$$S/N = \frac{\text{Energy transformed to perform the intended function}}{\text{Energy transformed to other than the intended function}}$$

Figure 7-1   Concept of S/N.

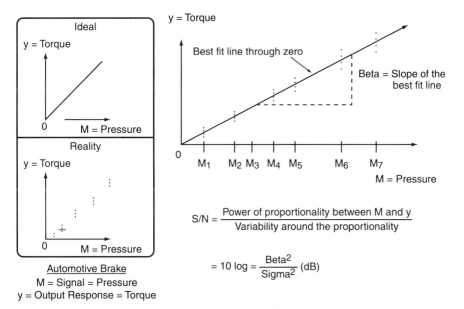

Figure 7-2   Calculation of S/N.

to reducing *variability* by half. Moreover, the QLF and the S/N are related measures. They are both measures of variability around the intended function. The S/N can easily be converted into the QLF so that quality improvement can be evaluated in monetary terms. An increase of S/N by 3 dB is equivalent to reducing the *loss* by half. Engineering objectives can be expressed in terms that are understood by management, and this facilitates communication throughout the organization as well as effective and efficient accomplishment of company objectives.

HOW TO EFFICIENTLY IMPROVE QUALITY AND COST EFFECTIVENESS (OFF-LINE QUALITY CONTROL)

The Taguchi Methods for Quality Engineering are a system for achieving company engineering objectives (i.e., for reducing loss and improving profit). Off-line quality improvement (i.e., accomplished during the design stage) includes the following three stages:

Stage 1. System Design: Conceptual design
Stage 2. Parameter Design: Robust design without a cost increase
Stage 3. Tolerance Design: Cost–quality trade-offs

Stage 1

System Design is the conceptual design stage in which scientific and engineering expertise is applied to develop new and original technologies. It takes creativity. The original conceptual design of the video cassette recorder (VCR), for example, was developed in the United States.

Quality Engineering techniques do not focus on this stage. Since it is not possible to study all potential systems (unless it is possible to perform computer simulations), Taguchi suggests that engineers select one, or just a few, concepts for development. Using the E-series of QFD matrices for concept selection can be quite useful to accomplish this engineering challenge.

Stage 2

Parameter Design is the stage at which a selected concept is optimized (quality is improved without a cost increase or S/N is maximized without a cost increase). Many variables can affect a system function. The variables need to be characterized from an engineering viewpoint. Some of the variables will be controllable (control factors). The engineer should set the nominal values of these variables based on known or anticipated nonlinear relationships between levels (settings) of control factors and selected performance measures. Others will be uncontrollable or too expensive to control (noise factors). These are the downstream variables that cause functional variation (see Figure 7-3).

The goals of Parameter Design do not including finding and removing these causes of variation or upgrading material and components to improve quality. The goals are to (1) find that combination of control factor settings that allow the system to achieve its ideal function and (2) remain insensitive to those variables that we cannot control (i.e., noise factors). This allows us to develop designs with high performance, stability and reliability. This stability of performance in the face of noise factors is called Robustness, Robust Design or Robust Quality. Improving robustness using Parameter Design allows the engineer to improve quality without increasing cost. It allows him or her to design products that are highly reliable under a wide range of conditions using less expensive materials, parts and methods.

The S/N evaluates the robustness of the function at selected control factor levels. The objective of Parameter Design is to reduce loss by

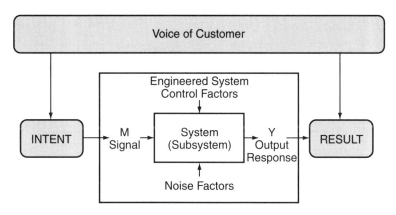

Parameter Design (Robust Design)
1. Design the experiment with Control Factors
2. Measure Output Response (y) as Signal (M) and selected noise factors are varied
3. Calculate S/N and find the combination of Control Factors that maximize the S/N

Figure 7-3 The engineered system and Parameter Design.

increasing robustness. Besides a quality improvement, there is still another benefit of Parameter Design.

Parameter Design provides opportunities to reduce the product and manufacturing costs. Once we have the means to improve quality without a cost increase, we can apply various cost reduction measures. Suppose the S/N was increased by 20 dB without a cost increase. Now we can apply some cost reduction measures that may result inreducing the S/N by 10 dB. We can do this because the quality is still better by 10 dB than it was in the original situation—and the cost is less at the same time! Therefore, we can increase sales while the cost is reduced. This increases the ability of a company to apply technology to compete economically and exploit its technological capability.

Stage 3

Tolerance Design is the stage at which variability can be further reduced by tightening tolerances. This requires the upgrading of the materials, parts and subcomponents of the system and generally increases cost. Tolerance Design attempts to further reduce variability with a minimum cost increase. it uses improvement measures for controlling some noise factors and/or compensation for the effects of noise factors, but it does this rationally. Rather than a costly upgrade of everything, it uses the QL1 to evaluate the impact of improvement measures with respect to cost, and a trade-off between cost and quality is found that balances cost and quality. Even though Parameter Design results in sufficient variability reduction, it is important to do the Tolerance Design to determine the tolerances and material grades. Note that if more robustness is achieved by Parameter Design, then less upgrading is necessary in Tolerance Design. In fact, it may happen that no upgrading is necessary in Tolerance Design.

The selection of quality characteristics that properly reflect the engineering function of a product or process is the most important and, perhaps, the most difficult task of the quality engineer. To evaluate robustness or improve existing product/process performance, what should we measure to accurately express the data and allow us to efficiently reduce variability (i.e., maximize the S/N)?

Characteristics such as yield, number of defects, percent defective, probability of failure, product life, mean time between failure, categorical sensory evaluation, and so on, are often used. Such characteristics do relate to economics and are needed to monitor and confirm the quality of products and processes. However, such characteristics do not directly express engineering function. They are often quite removed from the fundamental physics of product function and involve numerous extraneous interactions that can lead to problems downstream. Using these characteristics can compromise the efficiency and reliability of the experimentation.

For the upstream evaluation and improvement of products and processes, we need characteristics that accurately reflect engineering intent, those that are close to the basic energy transfer of the system and can be used to express performance, reliability and durability.

For example, a painting process may produce various types of defects such as voids, sags, and so on. Rather than counting the number of such defects or measuring yield as an output characteristic, a better strategy would be to recognize the intended function of the painting process. For example, one may decide to measure the relationship between the paint material flow rate (the signal) and the paint thickness (the output response). Then our objective would be to reduce the variability of the relationship (i.e., achieve a robust relationship). In other words, we are to create a process that can produce any paint thickness with the smallest variability.

This allows us to better study the variability of energy transfer without as many intervening variables. The use of data such as occurrence of defects, or yield, for reducing variability in the system is an inefficient strategy. Yield is an ineffective measurement because the sources of variability are compounded and the result of interactions of the symptoms of minimal robustness. Occurrence of defects is an inefficient measurement because it measures the symptoms of minimal robustness and leads only to fire fighting or problem solving without improving the robustness of the system. The selection of the best quality characteristic to express product/process function is dependent upon the particular case. It depends upon the engineer's knowledge of the particular field of technology and the cause-and-effect relationships therein.

Quality Engineering uses designed experiments to improve the S/N of product and process. The objectives of experimentation in Quality Engineering are significantly different from those of traditional Design of Experiments. Traditional experimentation is geared more toward meeting the objectives of the scientist working within the context of (original) research. The scientist seeks to discover and describe natural law or phenomena.

The objective of the practicing engineer is to achieve ideal product function, to reduce variability at the lowest cost, and to do it in the shortest development time. In learning to do this, the engineer accumulates technical knowledge for his or her company. The engineer needs methods that allow him or her to efficiently acquire technical information for improving quality while reducing cost. The engineer's activity is constrained by cost, time and other resources.

The job of the engineer is to specify product/process controllable parameters' nominal values, their tolerances, grades of materials, components, and so on, in such a way that cost and quality can be optimized, and this must be done as quickly and efficiently as possible. The resulting design creates products that reduce loss. In Parameter Design, an orthogonal array (a balanced experimental design matrix called a fractional factorial) should be used for examining controllable parameters, control factors (see Figure 7-3). The experiment is designed in terms of control factors and their levels. Then, for each combination of control factors, the output response is measured as the signal and selected noise(s) are varied so that an S/N can be calculated. Each control factor level is evaluated in terms of the S/N. The most robust design occurs when the S/N is maximized. Ultimately, the control factor levels that maximize the S/N are selected for use.

The objective is to achieve a robust function of the system, not necessarily to understand the mechanics of the phenomenon. While Taguchi Methods uses a balanced orthogonal array, as in traditional Design of Experiments, the application of these techniques and the philosophy behind them were developed for use by engineers to fit the realities of industrial design and production.

HOW TO COST EFFECTIVELY MONITOR AND MAINTAIN QUALITY (ON-LINE QUALITY CONTROL)

Even after quality has been improved and costs reduced using off-line methods, sources of variability still remain at the production level. These include variability in materials and purchased components, tool wear, machine fatigue, the drifting of process parameters, measurement error, and so on. To maintain the benefits of the off-line improvements, these sources of process variability must be controlled or compensated.

Taguchi has developed a system of on-line quality control methods that utilizes the QLF. It allows manufacturing engineers to design process control systems that minimize loss by monitoring the balance between cost and quality.

An on-line reduction in variability comes from the measurement and adjustment of equipment and process parameter levels. These procedures have associated costs, but Taguchi's On-Line system takes these costs into account. It determines how often a product characteristic or a process parameter should be measured and adjusted and what the optimum adjustment limits should be. This allows for economically informed decisions regarding production control measures in order to optimize process control systems.

This is quite different from the widespread strategy of control charting. Feedback and feedforward control are very important in Flexible Manufacturing and the design of automated processes. Automated control systems can produce expensive losses if they are not optimized. Feedback control measures a product or process parameter every n units and adjusts the process closer to the target using an adjustment factor. It thus compensates for the effects of variation rather than trying to remove the causes of variation. The frequency of measurement and the control limits are optimized as a function of the costs of measurement, adjustment and scrap/rework and the process capabilities of the manufacturing processes.

How often to measure and what control limits should be used can both be determined by Taguchi's On-Line methods. This on-line system includes methods for product control, prediction and correction, diagnosis and adjustment and preventative maintenance. What to measure, what to adjust, how much to adjust, how to improve adjustability and how to improve measurement are dealt with through off-line methods. With the application of off-line and on-line methods, control can be improved, monitored and maintained on the basis of an optimum balance between cost and quality.

The use and power of Taguchi's Robust Design concepts are well understood. They have been used by many organizations to design a wide variety of products throughout the world. Of course, the same is true of QFD. Historically, product development teams have used QFD methods or some of Taguchi's methods of Robust Design, but rarely have both methods been used together. Because of this we see

1. Products or services being developed that customers want *but that are not robust,* or
2. Products or services developed that are *very robust but do not meet customers' needs*

For maximum impact in the marketplace, both must be accomplished. That is, develop mature, robust technologies at the lowest cost in the shortest time all the while delivering on the customers' demands. The integration of QFD and Robust Design is not only smart, but virtually mandatory for companies that want to compete in today's fast paced competitive markets. Although these powerful tools have distinctly

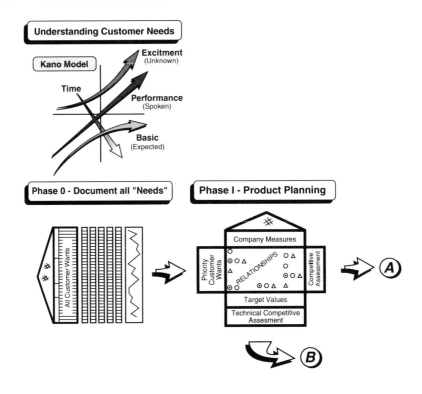

Illustrations by D.M. Verduyn

Reprinted with permission of the American Supplier Institute
© 996 American Supplier Institute

Figure 7-4   Integration of QFD and Robust Design flow chart.

different objectives, when properly integrated QFD and Robust Design can have a powerful synergistic effect.

The steps in Figures 7-4*a*, *b*, and *c* illustrate how QFD can be used as a framework to focus and prioritize customers' needs as well as to guide Robust Design efforts. The initial step is understanding your customers' needs better than your customers do. Many people insist, "Customers don't know what they want." They're wrong! Customers do know what they want, but they sometimes have difficulty articulating it. When you really understand the three different types of quality needs in the Kano Model and how to uncover these needs, your design team will be well on its way to understanding the customer needs better

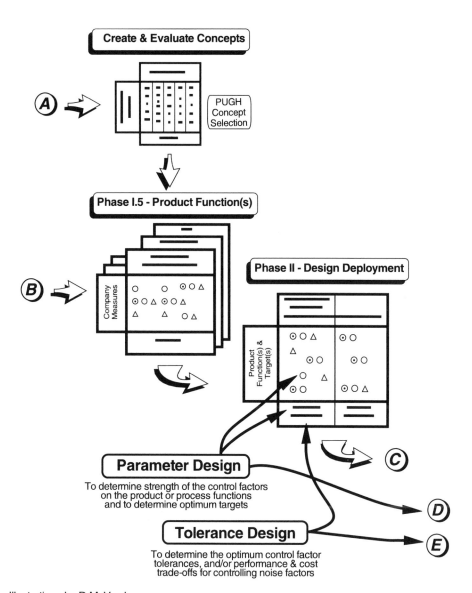

Illustrations by D.M. Verduyn

Figure 7-4   *(Continued)*

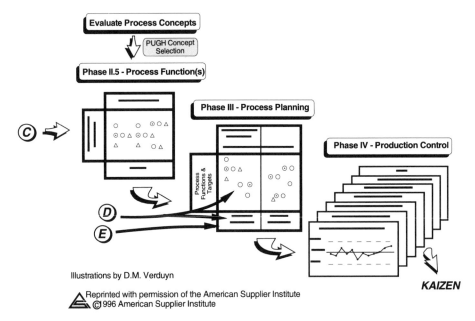

Figure 7-4 *(Continued)*

than the customer does. The Kano Model is quite useful in gaining insight into what motivates the customer (see Chapter 4).

The Pre-Planning Matrix in Figure 7-4 is simply a means to take the complex situation of many customer needs and, through the use of some predefined filters, reduce the number of customer wants that will be entered in the initial House of Quality. This shortcut helps to keep the size of the design project manageable to ensure project timeliness along with design team sanity! If the team tries to tackle too much in the QFD design effort, it can become overwhelmed and quickly lose team enthusiasm.

The object in Phase I (the Product Planning Matrix) is to translate the list of customer wants [from the Voice of the Customer (VOC)] to company (technical) measures that are objective and can be targeted. This phase also helps the team to understand competitive information. The complete chart (House of Quality) is then analyzed for priorities. These priorities are then deployed to one or more (depending on the complexity) Phase I.5 (Product Function Matrices) to develop the related product functions.

If alternative product design concepts are available and being considered, then the Pugh Concept Selection process provides an objective way of thoroughly evaluating these design alternatives. In addition, it often helps to synthesize the "best of all worlds," that is, come up with a new design that is better than all the initial alternatives. The Pugh technique compares alternative design concepts to a datum (or base) concept, typically the current design, using a complete set of evaluation criteria. The Pugh comparative analysis might include, for example, such criteria as ease of set-up and use, ease of obtaining superior results, weight and energy consumption. The objective of the Pugh technique is not necessarily to pick winners or losers but to gain insight and ideas from the many possible alternative concepts along with taking the subjectivity and biases out of the analysis.

This new, added phase that ties QFD to Robust Design is shown in the Phase I.5 Product Function Matrix. It is this matrix that translates the company or technical measures into product functions. The product function is something that, if done well, will often satisfy numerous company measures. In other words, company measures (the HOWs in Phase I) are typically indirect, downstream ways of evaluating how well the product is performing and accomplishing its job.

The key strategy of the product function is focusing the product development on efficient energy transformation. The product function is best defined in terms of energy transformation. The better the product is able to transform energy, the less lost energy there is to cause problems downstream. You have an ideal product function when there is perfect energy transformation. For example, the product function of a V-belt is to transfer mechanical power from the crankshaft to the water pump. The better the power transfer function is carried out, the less lost energy in the system causing problems such as noise, vibration, wear, and so on. By focusing the design effort on one product function, several company/technical measures are addressed simultaneously. This improves the efficiency of the design and concept development process, that is, "killing lots of birds (measures) with one stone (product function)."

The Phase II (Design Deployment) Matrix translates product functions and their related targets into a list of product design parameters (whose factors can be controlled) and noise factors (factors that cannot be controlled or are prohibitively expensive to control). Examples of the latter include the environment, variability in operating conditions, material, personnel skills, and so on. The noise factors are causes of variation and the control factors are the design parameters that will help achieve the robust product function. Robust Design is applied here in two stages. In the first stage, a Parameter Design study is conducted to evaluate the control factors against noise factors so as to determine (a) the control factor effects and (b) the optimum targets to achieve the robust product function. In the case of the earlier V-belt example, control factors such as belt material, belt thickness, belt cross-sectional angle, belt cord strength, belt cord elasticity and rubber hardness are tested against noise factors such as operating temperature, belt tension and dirt buildup.

Once the optimum targets for the control factors are determined, the proper tolerances around these target values are established through the application of the second stage, Tolerance Design.

If alternative process concepts are available, the Pugh Concept Selection technique will provide an objective and in-depth way of evaluating the process alternatives and often helps to arrive at the "best of all worlds," that is, a new process that is better than all of the initial alternatives. The Pugh technique compares alternative process concepts to a datum concept, usually the current process, using a complete set of evaluation criteria. The objective in using the Pugh technique is not to decide on winners or losers but to gain the greatest insight and number of ideas from among many alternative process concepts along with the removal of subjectivity and biases from the analysis.

Phase II.5 (Process Functions) ties QFD to Robust Design. The Phase II.5 matrix translates product control factors into process function(s). Similar to product functions, every manufacturing process has a function. The process function is something that, if done well, will produce products at their optimum targets with minimal variation. The process function strategy goes beyond the one-product-generation-at-a-time process optimization and focuses on a range of process applications, that is, a range of product targets. Optimizing the process over a range of product targets is developing the process technology for a family of today's and tomorrow's products. Once achieved, the process technology can simply be adjusted to produce a large variety of products. This improves the process development efficiency and shortens the overall product development cycle. Thus, fewer design resources are used to accomplish the same effect.

For example, a Computer Numerical Control (CNC) machining process is intended for the processing of shafts, discs and engine bores. The target values of these products vary somewhat, but all are addressed by the same CNC machining process function: to transfer CNC program dimensions into product dimensions, that is, the process function is to "create shape." If this process function is optimized, dimensions with minimal variation can be produced for a variety of products with varying targets.

The Phase III (Process Planning) matrix translates process functions into a list of process control factors and noise factors. Similar to Phase

II (Design Deployment), Robust Design is again applied in two stages, Parameter Design and Tolerance Design, to determine the optimum targets and tolerances for the process control factors. This will enable the process to produce a variety of products with minimal variation. These control factor optimum targets and tolerances then become guidelines for process control activities in Phase IV.

In Phase IV (Production Control) appropriate controls, practices and procedures, as determined by the findings in Phase III, are applied to ensure continuous process capability. The result is a knowledgeable and well-trained production workforce skilled in the appropriate application of gage repeatability and reproducibility studies, statistical quality control measures, preventive maintenance systems, mistake proofing, work-site operator aids and operator training.

8

QFD AND FAILURE MODE AND EFFECTS ANALYSIS

One of the many important reasons to use QFD is to obtain a clear statement at the very outset of a project of what issues, needs and requirements are important so that design efforts can be applied that are proportionate to the importance of the issues. Many organizations have in the past used a different method for establishing the critical issues in a product's design and thereby focusing the design efforts. The use of this method is often restricted to safety, reliability and durability issues, but it can be used beyond these typical applications. The method used is Failure Mode and Effects Analysis (FMEA) (also called Failure Mode, Effects and Criticality Analysis, or FMECA).

Firms involved in designing and manufacturing for the military routinely have contractual obligations to do FMECA studies early in their products and systems design cycle, typically during the concept/design stage. This is not a new approach to product design. The aerospace industry has been doing formal FMEAs since the mid-1960s. Manufacturers in the automotive industry have taken the FMEA studies approach the next natural step beyond the product design and have performed FMEA studies on the processes that will be producing the product. Since the mid-1980s, the U.S. Big Three automakers have each created different standards for their automotive product suppliers to meet. Recently these separate standards have been unified into one: QS-9000. Firms adhering to QS-9000 are required to create and use FMEA tables both during the product design process (called Product Design FMEA) and when designing the process that will manufacture that product (called Process FMEA).

An FMEA study generates a Risk Priority Number (RPN) for each failure mode. The higher the RPN is, the more serious the failure and the more important that it be addressed by the design effort. Usually the RPNs are arranged in descending order (except that any failures resulting in death or serious injury are placed at the top of the order, no matter what their actual RPN) and an arbitrary cutoff number is determined. All failures with RPNs higher than that cutoff number are addressed by the initial engineering efforts. Failures with lower RPNs may be addressed later if resources are available, or sometimes they are addressed as part of an engineering effort for a related failure that has a high RPN.

So the RPN that results from an FMEA study has a function similar to the importance ranking in an A1 matrix. It guides the resource allocation process so that the highest RPNs are given the greatest emphasis in the design effort. It is important that the high RPNs be reduced and preferably eliminated. It is this prioritization of RPNs that enables the failure items above the cutoff number to be moved from the FMEA study to the QFD matrix and then have appropriate importance rankings determined for each.

What the automotive, aerospace and military manufacturers of products and systems are currently doing to drive their design efforts through the use of FMEA studies can easily be extended to the design of

Services—for both (1) the services themselves and (2) the systems that support services delivery

Processes—when doing Business Process Re-engineering

Simply stated, the FMEA table is created by taking all known failure modes and evaluating them in three ways:

(a) *Severity (S)*. Criticality of the failure, if it does occur. (Is it just a nuisance or will it interrupt operations or cause major property damage, serious injury or even death?)

(b) *Occurrence (O)*. Likelihood that the failure will occur (e.g., frequency of occurrence over time or over a given number of opportunities).

(c) *Detection (D)*. Likelihood that the failure (or its preindicator) will be detected should it occur.

A typical evaluation system gives a number between 1 and 10 (with 10 being the worst case) for each of the three ways. The RPN is determined by multiplying together the three evaluations, that is, RPN = S × O × D. (See Tables 8-1 and 8-2 for guidance in determining the various rankings for Product Design FMEAs and Process FMEAs and see Figure 8-1 for an example of a Process FMEA Matrix.)

QFD and FMEA can assist each other in several ways. The values for severity, occurrence and detection are often determined subjectively by an internal, cross-functional group. Better results can be obtained by expanding this group to include both critical suppliers and customers, the same subgroups that should be included in a Voice of the Customer exercise. Suppliers can usually provide additional information on the frequency of failure and likelihood of failure detection. Customers can provide information on criticality and failure detection that an in-house group could not judge as well. Using a broader based group will give a better, more accurate knowledge base. In addition, this group may be able to finish the FMEA and then translate it into importance rankings for those House of Quality WANTs that were part of the FMEA (see Figures 8-2 and 8-3).

Organizations that have already done an FMEA for a similar product should still review it using a broader focus group to be sure that as much information as possible has been gleaned from both the suppliers and various customer groups. The weakness of many FMEA studies is that the entire activity was done by the firm's engineers and technicians. Obviously these people have a knowledge of the product being analyzed, but they may not have the full picture. To include a broader knowledge of the product's failure modes, it is essential that the team doing the FMEA study include suppliers, customers and especially the customer service personnel (both on-site service and the home office team that receives the customers' complaints). Various industry and government-funded groups (insurance, consumer, environment, etc.) may provide additional failure modes and frequency statistics.

If the company is adhering to QS-9000, the results of the process FMEA study should be correlated item by item to the Quality Plan to assure that every failure mode with a high RPN is addressed in the Quality Plan and is monitored to prevent occurrence. Particular failure modes that have a high severity, regardless of their occurrence or detection aspects, are also included in the group which has highest priority for improvement, mitigation or elimination.

In the case of the FMEA results and QFD, there is also a correlation. It is between the RPNs from the FMEA study and the durability, reliability and safety requirements for the product or service listed in the House of Quality. And, like the QFD House of Quality Matrix,

Table 8-1 Guide for product design FMEA evaluation

Severity

An evaluation of how critical the effect of a failure would be. The severity applies
to the effect only. *Note:* Severity can *only* be reduced or eliminated by design
changes. The following effects and ratings offer some ideas to evaluators. It is
often necessary to make modifications to fit a particular product or service.

Severity of Effect	Rating
Product/service becomes hazardous without warning.	10
Product/service becomes hazardous with some warning.	9
Product/service does not operate or does not perform its primary function.	8
Product/service is performing primary function but at a significantly diminished level such that the customer is dissatisfied.	7
Product/service is performing, but secondary function(s) are inoperable such that customer experiences some stress.	6
Product/service is performing, but secondary function(s) are at a significantly diminished level such that the customer is dissatisfied.	5
Aesthetic aspects do not conform to standards and nonconformance is detected by most customers.	4
Aesthetic aspects do not conform to standards and nonconformance is detected by the average customer.	3
Aesthetic aspects do not conform to standards and nonconformance is detected by knowledgeable/discriminating customers.	2
This failure has no effect from a severity standpoint.	1

Occurrence

An evaluation of the frequency of a failure occurring during the design life of the
product/service.

Suggested Evaluation Criteria

The team should agree on evaluation criteria and a ranking system that are
consistent even if modified for individual product analysis.

Probability of Failure	Possible Failure Rates	Ranking
Very high: failure almost inevitable	≥1 in 2	10
	1 in 3	9
High: repeated failures	1 in 8	8
	1 in 20	7
Moderate: occasional failures	1 in 80	6
	1 in 400	5
	1 in 2000	4
Low: relatively few failures	1 in 15,000	3
	1 in 150,000	2
Remote: failure unlikely	≤1 in 1,500,000	1

Table 8-1   *(Continued)*

Detection

Can current or proposed design controls defect design weakness or failure or preindications of failure? There are three types of design controls: (1) those that prevent the failure from ever occurring, (2) those that detect the cause or mechanism that leads to failure and (3) those that detect the failure mode. The *occurrence* rankings will be impacted by type 1 controls, provided they are integrated with overall design. The *detection* rankings will be based on type 2 or 3 controls, provided the stimulation (models, prototypes) reflect the design intent. *Note:* Detection ratings can *only* be reduced by improved design control (activities such as validation, simulation, verification and modeling)

Suggested Evaluation Criteria

The team should agree on evaluation criteria and a ranking system that are consistent even if modified for individual product analysis.

Detection	Criteria: Likelihood of Detection by Design Control	Ranking
Absolute uncertainty	Design control will not and/or cannot detect a potential cause/mechanism and subsequent failure mode or there is no design control	10
Very remote	Very remote chance design control will detect a potential cause/mechanism and subsequent failure mode	9
Remote	Remote chance design control will detect a potential cause/mechanism and subsequent failure mode	8
Very low	Very low chance design control will detect a potential cause/mechanism and subsequent failure mode	7
Low	Low chance design control will detect a potential cause/mechanism and subsequent failure mode	6
Moderate	Moderate chance design control will detect a potential cause/mechanism and subsequent failure mode	5
Moderately high	Moderately high chance design control will detect a potential cause/mechanism and subsequent failure mode	4
High	High chance design control will detect a potential cause/mechanism and subsequent failure mode	3
Very high	Very high chance design control will detect a potential cause/mechanism and subsequent failure mode	2
Almost certain	Design control will almost certainly detect a potential cause/mechanism and subsequent failure mode	1

Source: Potential Failure Mode and Effects Analysis (FMEA) Reference Manual. Copyright © 1990, Automotive Industry Action Group, Southfield, MI 48075.

Table 8-2   Guide for process FMEA evaluation

Severity

An assessment of the seriousness of the effect of the potential failure mode to the customer. The severity designation applies to the effect only. Depending on who the customer is that is affected by the failure mode, assessing the severity may lie outside the immediate process engineer's/team's field of experience or knowledge. In these cases, input from other knowledgeable persons must be sought. Prior to beginning the evaluation, the team should agree on evaluation criteria and a ranking system that are consistent even though in specific cases they may be modified for a particular process. Severity should be estimated on a scale from 1 to 10.

Severity of Effect	Ranking
May endanger machine or assembly operator. Very high severity ranking when a potential failure mode affects final product's operation and/or involves noncompliance with government regulation. Failure will occur without warning.	10
May endanger machine or assembly operator. Very high severity ranking when a potential failure mode affects final product's operation and/or involves noncompliance with government regulation. Failure will occur with warning.	9
Major disruption to production process. One hundred percent of product may have to be scrapped. Final product inoperable; loss of primary function. Customer very dissatisfied.	8
Minor disruption to production process. Product may have to be sorted and a portion (less than 100%) scrapped. End product operable, but at a reduced level of performance. Customer dissatisfied.	7
Minor disruption to production process. A portion (less than 100%) of the product may have to be scrapped (no sorting). End product is operable, but some comfort/convenience item(s) inoperable. Customers experience discomfort.	6
Minor disruption to production process. One hundred percent of product may have to be reworked. Product is operable, but some comfort/convenience item(s) operable at reduced level of performance. Customer experiences some dissatisfaction.	5
Minor disruption to production line. The product may have to be sorted and a portion (less than 100%) reworked. Visual aesthetics (i.e., fit and finish) or audible expectations (i.e., no squeak or rattle) are not met. Defect is noticed by most customers.	4
Minor disruption to production process. A portion (less than 100%) of the product may have to be reworked on-line but out-of-station. Fit and finish/squeak and rattle item does not conform. Defect noticed by average customer.	3
Minor disruption to production process. A portion (less than 100%) of the product may have to be reworked on-line and in-station. Fit and finish/squeak and rattle item does not conform. Defect noticed by discriminating customers.	2
No effect.	1

Table 8-2   *(Continued)*

Occurrence

Only occurrences resulting in the failure mode should be considered for this
ranking. Failure-detecting measures are not considered here. The following
occurrence ranking system should be used to ensure consistency. The possible
failure rates are based on the number of failures anticipated during the process
execution. If available, statistical data available from a similar process should be
used to determine the occurrence ranking. In other cases, a subjective assessment
can be made by utilizing the word descriptions under Probability of Failure. Cpk
values would come from capability studies of this or similar processes.

Suggested Evaluation Criteria

The team should agree on evaluation criteria and a ranking system that are
consistent even if modified for individual product analysis.

Probability of Failure	Possible Failure Rates	Cpk	Ranking
Very high: failure almost inevitable	≥1 in 2	<0.33	10
	1 in 3	≥0.33	9
High: generally associated with	1 in 8	≥0.51	8
processes similar to previous	1 in 20	≥0.67	7
processes that have often failed			
Moderate: generally associated with	1 in 80	≥0.83	6
processes similar to previous	1 in 400	≥1.00	5
processes that have experienced	1 in 2000	≥1.17	4
occasional failures, but not in			
major proportions			
Low: isolated failures associated with	1 in 15,000	≥1.33	3
similar processes			
Very low: only isolated failures	1 in 150,000	≥1.50	2
associated with almost identical			
processes			
Remote: failure unlikely; no failures	≤1 in 1,500,000	≥1.67	1
ever associated with almost			
indentical processes			

Detection

There are three types of process controls to consider:
1. Those that prevent the cause/mechanism or failure mode/effect from occurring
   or reduce their rate of occurrence
2. Those that detect the cause/mechanism and lead to corrective actions
3. Those that detect the failure mode

The preferred approach is to first use type 1 controls if possible; second, type 2
controls; and third, type 3 controls. The initial occurrence rankings will be
affected by the type 1 controls, provided they are integrated as part of the design
intent. The initial detection rankings will be based on the type 2 or 3 controls
used in current similar processes, provided the process being used is

Table 8-2   *(Continued)*

representative of process intent. Detection is an assessment of the probability that the proposed type 2 current process controls (such as listed in the eighth column from left of Figure 8-1, the example process FMEA) will detect a potential cause/mechanism (process weakness) or the probability that the proposed type 3 process controls will detect the subsequent failure mode before the part leaves the manufacturing operation or assembly location. A 1–10 scale is used (with 10 being the worst case). Assume the failure has occurred and then assess the capabilities of all current process controls to prevent shipment of the part having this failure mode or defect. Do not assume that the detection ranking is low because the occurrence is low (i.e., when control charts are used), but do assess the ability of process controls to detect low-frequency failure modes or prevent them from going further downstream in the process. Random quality checks are unlikely to detect the existence of an isolated defect and should not influence the detection ranking. Sampling done on a statistical basis is valid detection control method.

Suggested Evaluation Criteria

The team should agree on evaluation criteria and a ranking system that are consistent even if modified for individual process analysis.

Detection	Criteria[a]	Ranking
Almost impossible	No known control(s) available to detect failure mode	10
Very remote	Very remote likelihood current control(s) will detect failure mode	9
Remote	Remote likelihood current control(s) will detect failure mode	8
Very low	Very low likelihood current control(s) will detect failure mode	7
Low	Low likelihood current control(s) will detect failure mode	6
Moderate	Moderate likelihood current control(s) will detect failure mode	5
Moderately high	Moderately high likelihood current control(s) will detect failure mode	4
High	High likelihood current control(s) will detect failure mode	3
Very high	Very high likelihood current control(s) will detect failure mode	2
Almost certain	Current control(s) almost certain to detect failure mode; reliable detection controls known with similar processes	1

[a] Likelihood the existence of a defect will be detected by process controls before next or subsequent process or before part or components leaves the manufacturing or assembly location.
Source: Potential Failure Mode and Effects Analysis (FMEA) Reference Manual. Copyright © 1990, Automotive Industry Action Group, Southfield, MI 48075.

Item __Front Door L.H./H8HX-000-A__

Process Responsibility __Body Engrg./Assembly Operations__

Prepared By __J.Ford - X6521 - Assy.Ops__

Model Years(s)/Vehicle(s) __199X/Lion_4dr/Wagon__

Key Date _9X_03_01_ER_

FMEA Date (Orig.) _9X_05_17_ (Rev.) _9X_11_06_

Core Team __A. Tate Body Engrg., J. Smith-OC, R. James-Production, J. Jones-Maintenance__

Process Function / Requirements	Potential Failure Mode	Potential Effect(s) of Failure	S e v	C l a s s	Potential Cause(s)/ Mechanism(s) of Failure	O c c u r	Current Process Controls	D e t e c	R. P. N.	Recommended Action(s)	Responsibility & Target Completion Date	Action Results Actions Taken	S e v	O c c	D e t	R. P. N.
Manual application of wax inside door	Insufficient wax coverage over specified surface	Deteriorated life of door leading to: • Unsatisfactory appearance due to rust through paint over time • Impaired function of interior door hardware	7		Manually inserted spray head not inserted far enough	8	Visual check each hour-1/shift for film thickness (depth meter) and coverage	5	280	Add positive depth stop to sprayer	MFG Engrg 9X 10 15	Stop added, sprayer checked on line	7	2	5	70
To cover inner door, lower surfaces at minimum wax thickness to retard corrosion										Automate spraying	Mfg Engrg 9X 12 15	Rejected due to complexity of different doors on same line				
					Spray heads clogged - Viscosity too high - Temperature too low - Pressure too low	5	Test spray pattern at start-up and after idle periods, and preventative maintenance program to clean heads	3	105	Use Design of Experiments (DOE) on viscosity vs. temperature vs. pressure	Mfg Engrg 9X 10 01	Temp and press limits were determined and limit controls have been installed - control charts show process is in control Cpk=1.85	7	1	3	21
					Spray head deformed due to impact	2	Preventative maintenance programs to maintain head	2	28	None						
					Spray time insufficient	8	Operator instructions and lot sampling (10 doors / shift) to check for coverage of critical areas	7	392	Install spray timer	Maintenance 9X 09 15	Automatic spray timer installed - operator starts spray, timer controls shut-off control charts show process is in control Cpk=2.05	7	1	7	49

SAMPLE

Figure 8-1 Example of Process Design FMEA. [Reprinted with permission from the *APQP Manual* (Chrysler, Ford, General Motors Supplier Quality Requirements Task Force).]

POTENTIAL
FAILURE MODE AND EFFECTS ANALYSIS
(DESIGN FMEA)

FMEA Number ___1234___
Page _1_ of _1_
Prepared By _A. Tate X6412 Body Engr_

___ System
_X_ Subsystem 01.03/Body Closures
___ Component 01.03/Body Engineering

Design Responsibility _Body Engineering_   Key Date _9X 03 01 ER_   FMEA Date (Orig) _8X 03 22_ (Rev.) _8X 07 14_

Model Year(s)/Vehicle(s) _199X/Lion 4dr/Wagon_
Core Team _T.Fender-Car Product Dev, Childers-Manufacturing, J.Ford-Assy Ops (Dalton, Fraser, Henley Assembly Plants_

Item / Function	Potential Failure Mode	Potential Effect(s) of Failure	S e v	C l a s s	Potential Cause(s)/ Mechanism(s) of Failure	O c c u r	Current Design Controls	D e t e c	R. P. N.	Recommended Action(s)	Responsibility & Target Completion Date	Action Results				
												Actions Taken	S e v	O c c	D e t	R. P. N.
Front Door L.H. H8HX-0000-A • Ingress to and egress from vehicle • Occupant protection from weather, noise, and side impact • Support anchorage for door hardware including mirror, hinges, latch and window regulator • Provide proper surface for appearance items • Paint and soft trim	Corroded interior lower door panels	Deteriorated life of door leading to: • Unsatisfactory appearance due to rust through paint over time • Impaired function of interior door hardware	7		Upper edge of protective wax application specified for inner door panels too low	6	Vehicle general durability test vah. T-118 T-109 T-301	7	294	Add laboratory accelerated corrosion testing	A Tate-Body Engrg 8X 09 30	Based on test results (Test No. 1481) upper edge spec raised 125mm	7	2	2	28
					Insufficient wax thickness specified	4	Vehicle general durability testing- as above	7	196	Add laboratory accelerated corrosion testing Conduct Design of Experiments (DOE) on wax thickness	Combine w/test for wax upper edge verification A Tate Body Engrg 9X 01 15	Test results (Test No. 1481) show specified thickness is adequate. DOE shows 25% variation in specified thickness is acceptable.	7	2	2	28
					Inappropriate wax formulation specified	2	Physical and Chem Lab test- Report No.1265	2	28	None						
					Entrapped air prevents wax from entering corner/edge access	5	Design aid investigation with non-functioning spray head	8	280	Add team evaluation using production spray equipment and specified wax	Body Engrg & Assy Ops 8X 11 15	Based on test, 3 additional vent holes provided in affected areas	7	1	3	21
					Wax application plugs door drain holes	3	Laboratory test using "worst case" wax application and hole size	1	21	None						
					Insufficient room between panels for spray head access	4	Drawing evaluation of spray head access	4	112	Add team evaluation using design aid buck and spray head	Body Engrg & Assy Ops	Evaluation showed adequate access	7	1	1	7

SAMPLE

Figure 8-2 Example Product Design FMEA. [Reprinted with permission from the *APQP Manual* (Chrysler, Ford, General Motors Supplier Quality Requirements Task Force).]

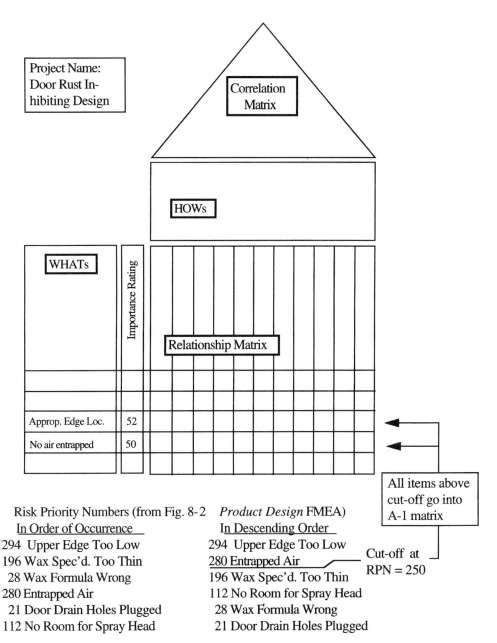

Project Name:
Door Rust In-
hibiting Design

Correlation
Matrix

HOWs

WHATs

Importance Rating

Relationship Matrix

Approp. Edge Loc.	52
No air entrapped	50

All items above
cut-off go into
A-1 matrix

Risk Priority Numbers (from Fig. 8-2 *Product Design* FMEA)

In Order of Occurrence	In Descending Order
294 Upper Edge Too Low	294 Upper Edge Too Low
196 Wax Spec'd. Too Thin	280 Entrapped Air
28 Wax Formula Wrong	196 Wax Spec'd. Too Thin
280 Entrapped Air	112 No Room for Spray Head
21 Door Drain Holes Plugged	28 Wax Formula Wrong
112 No Room for Spray Head	21 Door Drain Holes Plugged

Cut-off at
RPN = 250

Figure 8-3 House of Quality with entries from Product Design FMEA.

the FMEA tables need to be constantly updated as product design and process modifications are made so that in the next design cycle these documents will reflect the latest modifications. This will provide the new design team a running start and cut the time from concept to fully ramped production by an even greater margin than the previous cycle.

9

QFD AND THEORY
OF INVENTIVE
PROBLEM SOLVING

The Theory of Inventive Problem Solving (TRIZ in Russian, TIPS in English) has only recently become known in the United States. As will be explained later, TRIZ is a method that will help solve difficult technological problems though the identification and elimination of conflicts that are present in all engineered systems. Another feature unique to TRIZ is the ability of the method to predict how technical systems evolve over time. In the brief time that TRIZ has been practiced in North America and Europe, it has proved to be extremely powerful in generating elegant solutions to complicated paradoxical problems, the type of problems that product development teams are confronted with all the time. The rigor and graphical format of QFD have proved to be an effective method by which product development teams can organize and display the correlations between the product attributes. Product attributes that are negatively correlated are the inherent paradoxes, or system conflicts. A shortcoming of QFD is that the methodology does not possess any intrinsic ability to help the product development teams to resolve these conflicts. On the other hand, TRIZ is prolific at formulating conceptual solutions that can subsequently be evaluated for applicability within the context of the project.

The integration of the two techniques, TRIZ and QFD, makes perfect sense because of their complimentary nature. The following information outlines the significant elements of each methodology and will conclude with how QFD and TRIZ can be used as a powerful, integrated system to deploy the Voice of the Customer into unique and exciting new products and services.

THE LOGIC OF QFD

QFD organizes data into an L-type matrix, with customer requirements (the WHATs) and importance rankings in rows down the left side of the matrix and key product attributes as defined in engineered characteristics (the HOWs) in columns across the top of the matrix. In the cells where the rows and columns intersect, the strength of their relationship is recorded. Typically, relationships are categorized as one of three types: strong (indicated by a double circle), medium (single circle), or weak (triangle). If there is no relationship, then the row–column intersection (the cell) is left blank. Figure 9-1 gives an example in which column 3 has several strong relationships. Therefore, the column 3 engineered characteristic is quite important to meeting customer requirements. Column 7, on the other hand, only has one medium relationship, and if it is compared with column 3, it is obvious that column 3 is more important than column 7 because it positively impacts more customer requirements. In other words, it satisfies a larger number of customer wants. The technical characteristic called out in column 3 takes precedence should any trade-offs or conflicts arise between it and a less important HOW.

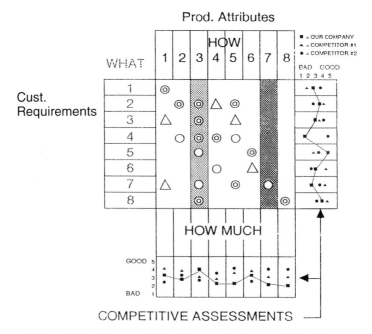

Figure 9-1 L-type matrix with several strong relationships.

Another valuable comparison on the House of Quality Matrix is the analysis that compares the engineered characteristics to each other. In the House of Quality, this comparison is done in the area referred to as "the Roof." This pairwise comparison of the engineered characteristics is done for a very specific reason, that is, to understand whether or not any of the system's technical attributes as depicted by the HOWs are in conflict. The importance of doing this analysis is that it promotes a holistic "systems" approach to graphically portraying and understanding the situation confronting the product development team. It is important because in every engineered system there are some elements of the system that are in harmony and complement each other, and there are others that are in conflict and contradict each other.

When doing a pairwise comparison between engineered characteristics, those that have a negative effect are denoted by a minus. Those with a strong negative effect are denoted by a circled minus. Similarly, a positive effect is indicated by a plus and a strong positive effect by a circled plus. In Figure 9-2 we see that there are strong negative effects between 4 and 8, between 1 and 7 and between 1 and 4.

SOLVING TECHNOLOGICAL CONTRADICTIONS

The potential problems caused by the number and nature of these system conflicts, as depicted in the roof analysis, have posed difficult paradoxical choices for product development teams. The typical response to resolving such a situation is to compromise, that is, to select parametric values at some point short of the ideal value for some or all of the conflicting parameters. For example, a pharmaceutical company manufacturing pain-relief medication might have to choose between the strength of the pain-relieving dosage versus how gentle the medication is on the stomach. Choosing parameters that provide not quite enough pain relief while allowing for some mild stomach upset is really an inferior strategy because neither customer want is really satisfied.

This scenario, however, is representative of how most technological conflicts are handled. The strategy is one of mitigating the negative impact on the customer, not complete customer satisfaction, that is, complete pain relief with no accompanying stomach upset. This begs the question "If a solution that satisfied the conflicting parameters was possible, how valuable would it be to the company in question?" It is not presumptuous to conclude that a company developing such a product would benefit greatly and would, for some period of time, enjoy an enviable market position. The innovative product creates a competi-

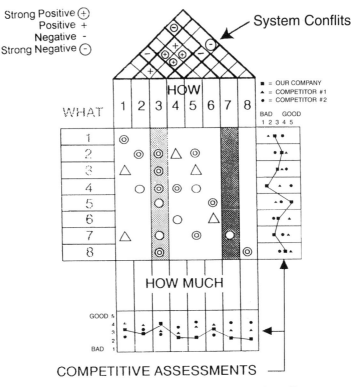

Figure 9-2 L-type matrix with strong negative effects.

tive gap or space, and because the space is largely unoccupied, competitors are left scrambling in headlong efforts to introduce their own competitive products. This is shown in Figure 9-3.

This phenomenon has repeated itself over and over again. Perhaps one of the most vivid recent examples has been the minivan introduced by Chrysler Corporation in the early 1980s. This vehicle exemplifies how a product, when engineered to answer contradictions such as ease of entry, interior space and ride dynamics, redefines the paradigm and creates the next-generation system. The Chrysler minivan successfully addressed the customers' paradoxical requirements and was a runaway best seller. For the next five years, other automobile companies hurriedly retooled their factories to manufacture a "me too" product. Chrysler, in the meantime, became known as the Minivan Company

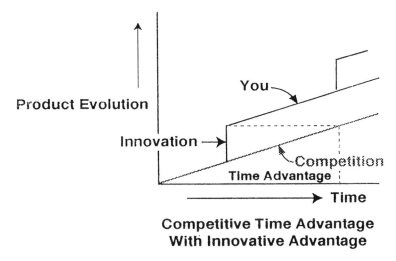

**Competitive Time Advantage
With Innovative Advantage**

Figure 9-3 Competitive time advantage with innovative advantage.

and successfully captured the market share as well as most of the profits. It is evident in today's competitive environment that innovative new products are the only real assurance of future prosperity. At Chrysler, the minivan was followed by the LH sedan, which proved to be another winner with its wide wheel base and low cab forward design. The LH also solved several conflicts between interior space and comfort versus fuel economy, power, handling, and so on. Chrysler, more than any other automobile company, exemplifies the importance of product innovation's impact on the bottom line.

QFD AND INNOVATION

Innovation and QFD should be a strategic element of any organization's long-term product planning process. While this seems obvious, in most organizations innovations occur rather haphazardly, if at all. The problem is that innovation is viewed as an unpredictable spontaneous event. Innovative ideas seem to spring from the murky subconscious of creative people in a serendipitous fashion. Any company that has introduced an innovative product will attest to the difficulty in repeating the scenario on a regular basis with subsequent products. Today it is axiomatic that, if the intervals of introducing new innovative products are spaced too far apart, the pressures of competition and the passing of time will erode the good will and reputation of the company. The market is always asking, "What have you done for me lately?" Today, organizations simply cannot defend their market position without regular infusions of innovative new products. Witness what happened to English motorcycles and what almost happened to the now reborn Harley Davidson Motorcycle Company.

If properly integrated into the organization's product development process, QFD will keep the organization "tuned" into the changing needs of its customers. QFD alone, however, will not generate innovative ideas. QFD will accurately portray the significance of the product's technical attributes and the resulting conflicts between these attributes. Enhancing the ability of the product development team to eliminate system conflicts, vis-à-vis innovative designs, is the compelling rationale for the integration of QFD and TRIZ. As mentioned earlier, resolving system contradictions through compromise is unsatisfactory, and groping for some stroke of innovative genius is risky and unpredictable. The promise of TRIZ is that it provides a structured scientific method for generating innovative ideas on demand.

INTRODUCTION

The Theory of Inventive Problem Solving, a structured scientific method, is more precisely and aptly described as the science of the evolution of technical systems. It was developed by a brilliant inventor named Genrich Altshuller while working in the patent office of the former Soviet Union. Altshuller was tasked by the Central Committee to study patents from around the world to extract and catalog strategic technologies and/or significant inventions the Soviets needed to be aware of. In the course of the study of some 400,000 inventions as depicted in patent descriptions, Altshuller noticed a fundamentally consistent approach used by the best inventors to solve problems.

At the heart of the best solutions, as described by the patents, existed an engineering conflict, or a "contradiction." The best inventions consistently solved conflicts without compromise. In other words, they eliminated the conflict through an innovative solution. The vast majority of inferior engineering solutions did not eliminate the contradictions, and these solutions were discarded. Upon closer examination and classification of innovative solutions, natural patterns of solutions started to emerge. Altshuller had discovered that when an engineering system was reduced to reveal the essential system contradictions, inventive solutions eliminated the contradictions completely. Furthermore, Altshuller noticed that the same inventive solutions appeared repeatedly at different points in time and in different countries. This startling discovery provided the fundamental rationale to expand the study of inventions to ascertain what other existing natural patterns were waiting to be discovered within the patent data base.

To date, 1,500,000 patents have been systematically examined, and a coherent set of Inventive Principles for the elimination of engineering contradictions have been discovered and cataloged. The extensive patent analysis has also revealed an explicit construct detailing systematic patterns of the evolution of engineering systems. The Inventive Principles and the Laws of Engineering Systems Evolution have been integrated with a rigorous algorithm that is the chief tool used within TRIZ to extrapolate and operationalize the innovations contained in the patent knowledge base.

BASIS FOR TRIZ

Underlying the unified set theory of TRIZ is the inexorable movement of technical systems toward the Ideal System. The Ideal System, in theory, is a state where the mechanism is absent but the function of the mechanism is present. The Ideal System therefore provides useful function only. Ideality is defined as the ratio of useful function versus harmful function. In a macro sense, technical systems move toward the state of ideality even if it is only theoretically possible to attain it in its ultimate form.

Emanating from the movement to Ideality are the laws of the evolution of technical systems. To be more accurate, these are not objective laws, as in physics or chemistry, but rather a consistent pattern of evolutionary phenomena that are manifest in technical systems. The significance of these evolutionary patterns to manufacturing companies lies in the strategic product decisions made by the organization. Product decisions running counter to the natural patterns spell potentially severe, if not disastrous, consequences to the organization.

LAWS OF THE EVOLUTION OF ENGINEERED SYSTEMS

Within the patent data base, Altshuller traced the chronology of patents and, as a result, formulated the theoretical basis for TRIZ, which is now referred to as the Laws of Engineering Systems Evolution. Within TRIZ, every significant improvement in a system is considered as evolutionary and a movement toward the Ideal System. There are eight

Laws of Evolution and approximately 200 sublines of more specific phenomenan occurring in the evolution of engineered systems. The eight laws are as follows:

1. *EMERGENCE OF A NEW SYSTEM.* A new technical system emerges when there is (a) a need and (b) the functional ability to meet that need. In other words, "necessity is the mother of inventions." In the common vernacular, this is called a New Paradigm. Initially, new technical systems leave much room for improvement. The increase in Ideality of the system then proceeds along predictable evolutionary paths through development, maturity, decline and, ultimately, replacement by the next paradigm.

2. *INCREASING IDEALITY.* After emergence, systems improve in their ability to provide more useful function for less cost. This is due in part to the incorporation of the resources in and around the system and/or the elimination of some of the contradictions in the system.

3. *UNEVEN DEVELOPMENT OF SUBSYSTEMS.* In time, certain subsystems become more technically advanced than others in the supersystem. This creates subsystem-to-subsystem dissonance, and the lagging subsystem paces the development of the supersystem. Engineering and product development teams will work at improving the lagging subsystem to reduce the dissonance in the supersystem.

4. *DYNAMICITY.* As systems evolve, they become more dynamic. This is manifest when the system is able to perform more function, that is, "it slices and dices," or it has increasing degrees of freedom. The transition from a one-way hinge to a ball in socket is an example of additional degrees of freedom.

5. *TRANSITION TO BI- AND POLY-SYSTEMS.* One technical system will combine with another to provide a new hybrid supersystem. For example, the facsimile (fax) machine has evolved from a single-purpose piece of equipment into a polysystem that performs its original function plus additional functions such as printing, copying and scanning.

6. *HARMONIZATION OF RHYTHMS.* As systems approach Ideality, the dissonance between subsystems is reduced to a point that further system evolution is incremental. At this point, a period of stability in the design and functions performed is reached. Examples of this are power tools, bicycles and some home appliances.

7. *TRANSITION TO MICROLEVELS AND INCREASED USE OF ENERGY FIELDS.* As time passes, systems become more efficient in the transfer of energy into useful function. The use of molecular and energy fields are incorporated, and oftentimes the system will shrink in size. The microwave oven and the cellular phone are examples of this law.

8. *INCREASING LEVELS OF AUTOMATION.* In its ultimate form, the system becomes totally autonomous and self-regulating. The need for human interaction is minimized if not eliminated entirely. Automatic pilots, inertial guidance systems and smart suspensions are examples of autonomous systems.

The significance of these eight laws is that any system can be examined and mapped to a life-cycle curve. In other words, it is possible to determine which of the lines of evolution have manifested themselves and, thereby, it is possible to predict with a high degree of certainty how the system will evolve in the future.

MOVEMENT TOWARD IDEALITY

The notion of the Ideal System provides a powerful springboard to solving technical problems because the idea itself helps to overcome Psychological Inertia, a condition described by Joel Barker as Paradigm Paralysis. Within the lexicon of TRIZ, this mental confinement is termed Psychological Inertia. When the Ideal System is described, the attainment of it is most often blocked by stubborn engineering contradictions that have heretofore eluded elimination through innovative solutions. Within TRIZ, there are methods to systematically reveal all of the engineering conflicts and to search in an efficient and scientific way for solutions that eliminate these conflict(s). It is the pursuit of the Ideal System that acts as the fundamental rationale in seeking uncompromising inventive solutions. The attainment of the Ideal System, then, is a primary subverter of Psychological Inertia.

ALGORITHM FOR INVENTIVE PROBLEM SOLVING (ARIZ)

The principal tool within TRIZ for elimination of Engineering Contradictions is ARIZ. The objective of ARIZ is to systematically break down the problem(s) within any system into a smaller set of so-called miniproblems and to find elegant solutions utilizing the principles uncovered within the patent data base as well as the resources within the system. It is possible, therefore, to resolve in a totally acceptable fashion the Engineering Contradiction(s).

The Algorithm for Inventive Problem Solving was developed to provide problem solvers with a consistent way of analyzing a problem to reveal all of its facets. It accomplishes this by systematically decomposing the elements of a problem from the supersystem to the system, to the subsystem, to the component and to molecular levels, if required. In this process, the original problem formulated by ARIZ will oftentimes migrate to an unsuspected area of the system that provides an easier solution. In other words, ARIZ will often identify a totally different problem than was originally anticipated and, hence, a totally different and more elegant solution. The advantage of utilizing ARIZ is that, being derived from the laws of evolution of technical systems, it will generate concepts of solutions that modify the system according to those natural patterns of evolution. Running counter to the movement toward Ideality is futile in the long run because someone will surely introduce a better system, and the marketplace will unquestionably gravitate toward it.

CONTRADICTIONS

Another fundamental approach within TRIZ is the acknowledgment and straightforward attack on contradictions. There are two primary types of contradictions identified in TRIZ: Engineering Contradictions and Physical Contradictions. Contradictions, also called Engineering Conflicts or Paradoxes, pose the primary challenge to making meaningful improvements in the system. The typical response to such contradictions is to make some sort of compromise. Note the words of the American theologian Tryon Edwards: "Compromise is but the sacrifice of one right or good in the hope of retaining another, too often ending in the loss of both." (*The New Dictionary of Thoughts,* Catchogue, N.Y.: Buccaneer Books, 1957)

ENGINEERING CONTRADICTIONS

Engineering Contradictions are defined as the direct opposition of elements or parameters within the system. For example, in a computer hard disk drive, customers want fast access time and lots of storage capacity. In reality, given the current technology, the access time to retrieve information is in conflict with the amount of electronic data stored on the hard disk. In other words, the more information stored on the disk drive, the longer it takes to retrieve any specific piece of data. Other well-known contradictions include conflicts such as weight versus strength, or power versus heat, and so on.

The typical approach to solve the contradictions is to ignore them or, as stated earlier, to accept compromise. This is unsatisfactory because neither parameter is satisfied and the system is not improved in a meaningful way. Within the database of patents, Altshuller discovered that in the most elegant inventions contradictions were not compromised but eliminated. As the contradictions and the resulting solutions were cataloged, it was startling that a relatively small number (40) of the Inventive Principles solved many of the Engineering Contradictions. It was also surprising as to the applicability of the principles in eliminating contradictions in widely different engineering disciplines. Some of these principles are summarized in Figure 9-4.

Having discovered the Inventive Principles for the elimination of Engineering Contradictions, Altshuller devised a clever way of mapping the conflict to the Inventive Principles. To provide a means by which the appropriate principle is called out, Altshuller constructed a 39 × 39 Engineering Parameter Table. A portion of the Parameter Table is shown in Figure 9-5.

The table is deceptively simple to use. One has merely to select a parameter to improve (in the rows) and match it to one or more parameters that are undesired (in the columns). At the intersection (cell) of the two conflicting engineering parameters, there is a series of numbers that correspond to the appropriate Inventive Principle to be considered to eliminate the contradiction. The principles that are called out in each cell are derived from the patent data base and provide the basis for an inventive solution.

PHYSICAL CONTRADICTIONS

Physical Contradictions occur when a single parameter or element in the system must be in self-opposing states simultaneously. For example, in the electroplating process, to optimize the plating deposition, the temperature of the solution bath needs to be hot. But when the solution bath temperature is hot, the shelf life of the bath is reduced. The Physical Contradiction is solution bath temperature because it needs

1. Segmentation
    a. Divide an object into independent parts
    b. Make an object sectional (or into modules)
2. Extraction
    a. Remove or separate a problematic component or property of an object
    b. Extract the harmful product of a system
3. Local Quality
    a. Modify the structure of an object or its environment from a homogeneous to a heterogeneous state
    b. Place each subsystem in a condition favorable for its function
4. Asymmetry
    a. Replace a symmetrical form with an asymmetrical form
    b. In an asymmetrical product, increase the degree of asymmetry

13. Inversion
    a. Implement an opposite action
    b. Turn the object upside down, inside out, etc.
14. Spheroidality
    a. Replace linear parts or flat surfaces with curved ones, or cubical or spherical shapes
    b. Replace linear motion with rotating motion
15. Dynamicity
    a. Make characteristics or elements of an object adjust automatically for optimal performance
    b. Make objects moveable or interchangeable

39. Inert Environment
    a. Replace normal environment with inert one
    b. Carry out process in vacuum
40. Composite Materials
    a. Replace homogeneous materials with composite materials

Figure 9-4   Some of The 40 Inventive Principles.

to be hot for plating and, at the same time, it needs to be cold for shelf life. Altshuller discovered that oftentimes a Physical Contradiction was present as a controlling parameter between two Engineering Contradictions. Therefore, if a solution was to resolve, the seemingly paradoxical situation could be found by resolving the Physical Contradiction; the Engineering Contradictions would be resolved in an elegant manner. If, as posed above, the solution bath could be made to be hot and cold

Feature to Change (Parameter) \ Undesired Result (Conflict)	1 Weight of moving object	2 Weight of non-moving object	3 Length of moving object	4 Length of non-moving object	5 Area of moving object	6 Area of non-moving object	7 Volume of moving object	8 Volume of non-moving object	9 Speed	10 Force	11 Tension, pressure	12 Shape	13 Stability of object
1 Weight of moving object			15,8, 29,34		29,17 38,34		29,2, 40,28		2,8, 15,38	8,10, 18,37	10,36, 37,40	10,14, 35,40	1,35, 19,39
2 Weight of non-moving object				10,1, 29,35		35,30 13,2		5,35, 14,2		8,10, 19,35	13,29 10,18	13,10, 29,14	26,39, 1,40
3 Length of moving object	8,15, 29,34				15,17 4		7,17, 4,35		13,4, 8	17,10, 4	1,8, 35	1,8, 10,29	1,8, 15,34
4 Length of non-moving object		35,28 40,29				17,7, 10,40		35,8, 2,14		28,10	1,14, 35	13,14, 15,7	39,37, 35
5 Area of moving object	2,17, 29,4		14,15 18,4				7,14, 17,4		29,30 4,34	19,30 35,2	10,15, 36,28	5,34, 29,4	11,2, 13,39
6 Area of non-moving object		30,2, 14,18		26,7, 9,39						1,18, 35,36	10,15 36,37		2,38
7 Volume of moving object	2,26, 29,40		1,7, 4,35		1,7, 4,17				29,4, 38,34	15,35 36,37	6,35, 36,37	1,15, 29,4	28,10, 1,39
8 Volume of non-moving object		35,10, 19,14	19,14	35,8, 2,14						2,18, 37	24,35	7,2, 35	34,28, 35,40
9 Speed	2,28, 13,38		13,14, 8		29,30 34		7,29, 34			13,28, 15,19	6,18, 38,40	35,15, 18,34	1,18
10 Force	8,1, 37,18	18,13, 1,28	17,19, 9,36	28,10	19,10, 15	1,18, 36,37	15,9, 12,37	2,36, 18,37	13,28, 15,12		18,21, 11	10,35, 40,34	35,10, 21
11 Tension, pressure	10,36, 37,40	13,29, 10,18	35,10, 36	35,1, 14,16	10,15, 36,25	10,15, 35,37	6,35, 10	35,24	6,35, 36	36,35, 21		35,4, 1, 0	35,33, 2,40
12 Shape	8,10, 29,40	15,10, 26,3	29,34, 5,4	13,14, 10,7	5,34, 4,10		14,4, 15,22	7,2, 35	35,15, 34,18	35,10, 37,40	34,15, 10,14		33,1, 18,4
13 Stability of object	21,35, 2,39	26,39, 1,40	13,15, 1,28	37	2,11, 13	39	28,10, 19,39	34,28, 35,40	33,15, 28,18	10,35, 21,16	2,35, 40	22,1, 18,4	
14 Strength	1,8, 40,15	40,26, 27,1	1,15, 8,35	15,14, 28,26	3,34, 40,29	9,40, 28	10,15, 14,7	9,14, 17,15	8,13, 26,14	10,18, 3,14	10,3, 18,40	10,30, 35,40	13,17, 35
15 Durability of moving object	19,5, 34,31		2, 19, 9		3,17, 19		10,2, 19,30		3, 35, 5	19,2, 16	19,3, 27	14,26, 28,25	13,3, 35
16 Durability of non-moving object		6,27, 19,16		1,10, 35				35,34, 38					39,3, 35,23
17 Temperature	36,22, 6,38	22,35, 32	15,19, 9	15,19, 9	3,35, 39,18	35,38	34,39, 40,18	35,6, 4	2,28, 36,30	35,10, 3,21	35,39, 19,2	14,22, 19,32	1,35, 32
18 Brightness	19,1 32	2,35, 32	19,32, 16		19,32, 26		2,13, 10		10,13, 19	26,19, 6		32,30	32,3, 27
19 Energy spent by moving object	12,18, 28,31		12,28		15,19, 25		35,13, 18		8,15, 35	16,26, 21,2	23,14, 25	12,2, 29	19,13, 17,24
20 Energy spent by non-moving object		19,9, 6,27									36,37		27,4, 29,18

Figure 9-5   A partial Engineering Parameter Table.

at the same time, then the conflict between plating efficiency and shelf life is resolved.

The elimination of Physical Contradictions was accomplished, as Altshuller discovered, through the incorporation of a totally different strategy and a different set of principles. Physical Contradictions in the most elegant inventions are eliminated through the incorporation of the Separation Principles. Some of the Separation Principles are separation in time, separation in space, separation in scale and separation between the component and the system. As an example, the curvature (camber) of an airplane wing needs to be changed from takeoff to cruise to landing. This change is accomplished through the invention of flaps that extend and retract as necessitated by the situation. In other words, wing camber is large and small at different points in time. Bifocal lenses illustrate the principle of separation in space. The lenses are thin and thick at different places. The incorporation of this principle provides the most elegant solution to the problem of making the plating bath temperature both hot and cold. Can you formulate a solution utilizing the principle of separation in space? The answer is provided at the end of the chapter.

Within the hierarchy of preferred solutions, the utilization of the Separation Principles to resolve Physical Contradictions generally provides more elegant solutions. In the absence of a Physical Contradiction, the Engineering Contradiction can be eliminated through the application of one or several of the Inventive Principles.

SUBSTANCE FIELD ANALYSIS

Substance Field Analysis is to TRIZ what algebra is to mathematics. For example, in algebra we use abstract symbols ($a, b, x,$ etc.) to describe the structure of a problem. The very fact that a problem is depicted in an abstract format makes it easier to use the rules of algebra to solve the equation. Algebraic notation is simply a convenient way to describe the construct of the problem as well as the steps to the solution. The real values that define the parameters of the problem are not nearly as important as the actual steps used in solving the problem. In fact, using the real values will obscure the understanding of how the solution was derived. Substance Field Analysis, abbreviated S-Field Analysis, is a way to describe how various substances (elements of the system) and fields (electrical, magnetic, mechanical, thermal, electromagnetic, etc.) interact with each other to produce an effect. In the analysis of patents, when the various problems were described using

S-Field nomenclature, Altshuller was able to compare S-Field diagrams before (preinvention) and after (postinvention). Just as in algebra, the S-Field transformations exhibited identical structural modifications irrespective of the engineering domain of the problem. Just as the steps to solve a quadratic equation do not depend on the numeric values used, the elimination of engineering and physical contradictions are not dependent on the actual substances and fields that compromise the system. Within the data base of patents, there are 76 standard S-Field solutions. One has to keep in mind that the catalogued S-Field transformations generated elegant solutions eliminating numerous contradictions. In other words, the new S-Field structure represents a significant improvement in the system, one that substantially increases the Ideality ratio.

Within the context of product development, this can be of great strategic importance because the S-Field transformations lead to "next-generation" systems. Next-generation systems typically define the next paradigm and thereby create the competitive space occupied solely by one's self.

OTHER TOOLS OF TRIZ

So far, we have examined the fundamental building blocks and logic of TRIZ:

1. Technical Systems evolve over time to higher states of Ideality. This movement is inexorable, traceable and predictable.
2. The evolution of Technical Systems is explained and chronologically fixed through the Laws of Engineering Systems Evolution.
3. The movement toward Ideality results primarily through the elimination of engineering conflicts.
4. TRIZ incorporates a number of operational tools that provide systematic and scientific ways of prodding system improvements.
5. These tools have been discovered through a massive study of the world's best inventions.
6. The tools of TRIZ can be used to solve chronic problems in today's systems as well as to predict and subsequently solve problems in tomorrow's systems.
7. The primary logic of TRIZ is embodied in ARIZ.
8. The Algorithm for Inventive Problem Solving incorporates numerous subtools and techniques, such as Inventive Principles,

Separation Principles, and Substance Field Analysis, to overcome Psychological Inertia and to discover innovative possibilities heretofore beyond the mental reach of the engineer.

In addition to the above-mentioned elements of TRIZ, there are many others incorporated in the technique that are beyond the scope of this chapter and book. Some of these are the following:

(A) *TECHNOLOGICAL FORECASTING.* As discussed earlier, Technical Systems evolve to become functionally superior to their predecessors. Within TRIZ the point of evolution can be mapped to technology life-cycle S curves. The resulting analysis can then predict with high probability how the system can and should evolve to increase the system's functionality and, hence, Ideality.

(B) *SUBVERSION ANALYSIS (SA).* Subversion analysis is analogous to Failure Mode and Effects Analysis (FMEA) with one clearly distinct difference. In FMEA the question is "How can this system fail?" Subversion Analysis asks the question "How can I make this system fail?" The resulting answers lead to rich opportunities for system improvements that embody countermeasures to potential failures derived from a proactive instead of reactive mindset.

(C) *KNOWLEDGE BASE OF EFFECTS.* Embodied within TRIZ are 250 physical effects, 120 chemical effects and 50 geometrical effects. Often problems defy solutions through conventional approaches. The above-mentioned effects provide another powerful weapon within the TRIZ arsenal to utilize little known and seldom used elements of physics, chemistry and structures to enhance and broaden the solution space.

(D) *RESOURCES.* Every system has readily available resources that can be used to improve the Ideality of the system. The resources can be substances, energy, waste by-products, space, differentials within the system, functions, and so on. There are also resources available in the surrounding environment or within the supersystem. TRIZ will rigorously explore all of the available resources and utilize them to the extent possible to improve the system.

(E) *SMART LITTLE PEOPLE.* Another somewhat unique and curious tool of TRIZ is the so-called Smart Little People (SLP) Method. The objective of the SLP Method is to help break psychological inertia through visualization and role play. When using the SLP Method, people actually act out what is happening; that is, they mimic the conflicts of the problem and they mimic the ideal solution. Stepping back and observing the dynamics of the situation and the ideal solution have provided obvious solutions heretofore unimagined.

(F) *CONFLICT INTENSIFICATION (CI)*. Conflict intensification is used as a step in ARIZ to prevent the problem solver from accepting compromising solutions. Conflict intensification will also suggest solutions visible only under an intensified condition and also helps to break Psychological Inertia.

QFD, TRIZ AND ROBUST DESIGN

One of the Laws of Engineering Systems Evolution, law 5, transition to bi- and polysystems, outlines how single systems combine with other single systems to form new hybrid supersystems. In a sense, that is happening with QFD, TRIZ and Taguchi's Robust Design. Each methodology by itself has unique strengths and is designed to fulfill a particular function. QFD, for example, provides the ability to translate the Voice of the Customer into technical specifications; TRIZ analyzes a system and defines all of the inner system conflicts and suggests how to eliminate conflict through innovation; and Taguchi's Robust Design is very powerful in providing engineers the ability to make generic functions within a technology robust. Robust technologies naturally produce robust products. As one analyzes the three techniques, it becomes readily apparent that there is a synergy and a seamless continuity between the three. The three methodologies support each other like the three legs of a stool and provide an all-encompassing completeness heretofore not available to U.S. companies.

TRIADS, a service mark of the Renaissance Leadership Institute of Oregon House, California, is the name given to the three integrated tools. In the opinion of the author, TRIADS represents the next paradigm in new product development and technological problem solving. Through TRIADS, one is able to understand and translate customer requirements, evaluate those requirements for conflicts, access a huge knowledge base of inventions to eliminate the conflicts by using innovation and, finally, take the chosen innovative concept to a state of robustness through parameter design optimization. Figure 9-6 illustrates the integration of the three methods.

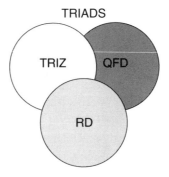

Figure 9-6   TRIADS.

CONCLUSION

We are at the forefront of a revolutionary and fundamentally different way of developing new products and solving paradoxical engineering problems. Innovation can and must become a planned phenomenon if we are to gain and retain competitive advantage. Joel Barker, the well-known futurist, warns against thinking "that the future is merely an extension of the past." Barker explains that when a paradigm shifts "everyone goes back to zero." In other words, the processes one may have used to solve problems or to generate innovations in the past will not be sufficiently effective when the rules have changed. We are at the beginning of such a paradigm shift. Organizations simply cannot leave the process of generating innovative ideas to trial and error or to sporadic flashes of brilliance if they are to remain competitive.

Genrich Altshuller, in his book *Creativity as an Exact Science* (1988, p. 273), states:

> And so, inventions, even those of the very highest level can also be made according to formula. Many of these formulae are already known and their application is studied with success. In the next decades, the solution of inventive problems will be converted into an exact science on the development of technical systems. . . . evidently the possibilities of controlling the process of thought are boundless. They cannot be exhausted because reason, the greatest instrument of knowing and transforming the world, is also capable of transforming itself. Who can say that there is a limit to the process of humanization of man? As long as

man exists, the control of this force will be improved. We are only at the start of a long road.

The answer to the plating problem stated earlier in this chapter is to heat the parts. The plating solution around the parts is warm for optimal plating but cool in the rest of the bath for optimal shelf life.

10

QFD FOR
SERVICE INDUSTRIES

Quality Function Deployment was initially created to assist in focusing the design process to develop products that satisfy customers. Then, service industries "discovered" QFD and its ability to help in designing services. Some early applications (in 1981) of applying QFD to service organizations in Japan were for a shopping mall, a sports complex and a variety retail store. In more recent Japanese activity, QFD has been integrated with reliability and quality circle activities in hotels, shopping centers and hospitals.

Since 1983, a number of leading North American firms have discovered this powerful approach and are using it with cross-functional teams and concurrent engineering to improve both products and services as well as to improve the design and development process itself. The author used QFD in 1985 to develop his Japanese translation business, Japan Business Consultants, and saw revenues increase 285% the first year, 150% the second year and 215% the third year. (The original study has been updated and is presented as a case study later in this chapter.) QFD was an integral part of Florida Power & Light's successful bid to become the first non-Japanese Deming Prize recipient in 1990. QFD has also been successfully applied in the U.S. health care industry since 1991 at the University of Michigan Medical Center and the Medical Center of Central Massachusetts.

Since 1990, the author has consulted with other service organizations in distribution, education, personnel, finance, health care, repair and retail businesses. QFD has provided a structure for assuring quality and customer satisfaction in the otherwise fuzzy and intangible world of service.

WHY QFD FOR SERVICES?

Increasing economic pressures from competition, government, and rapidly changing technology have forced companies to ask fewer employees, often with fewer resources, to accomplish more. Internal company services such as personnel, accounting, information management, and so on, are no longer ancillary activities but have become critical processes in assuring both internal and external customer satisfaction and in achieving overall organizational objectives. How can this be accomplished with ever-diminishing financial, time and human resources?

What about service-oriented businesses? For example, there are mounting pressures for health care reform that will undoubtedly mean fewer people with fewer resources doing more for more customers. How will they assure that the quality of health care will not suffer?

What about small business? In *Liberation Management* Peters (1992, p. 142) describes his personal view of the consulting firm McKinsey & Company as an organization with consultants (professionals) and support staff (second-class citizens). As long as the staff remain at McKinsey, they will never rise to top positions (partnership). For these support staff to become first-class citizens, they must join an organization that specializes in support activities (research, duplicating services, desktop publishing, transcription, etc.), where they can be "professionals" in their own right. Peters sees a North America proliferating with service firms electronically linked to their customers. What will be the basis of these linkages?

Why look to QFD to address the problems of services? What can QFD do that is not already being done by traditional quality systems? In understanding QFD, it is helpful to understand the differences between modern and traditional quality systems.

"NOTHING WRONG" DOES NOT EQUAL "EVERYTHING RIGHT"

Traditional Quality Systems

Traditional approaches to assuring service quality often focus on work standards, automation to eliminate people or, in more enlightened organizations, quality improvement teams to train and empower employees to solve problems.

As manufacturers are finding out, however, consistency and absence of problems is not a competitive advantage when only good players are left. For example, in the automobile industry, despite the celebrated narrowing of the "quality" (read as "fit and finish") gap between U.S. and Japanese makers, Japanese cars still predominate in the award of top honors in the J.D. Powers' Survey of New Car Quality.

Modern Quality Systems

QFD is quite different from traditional quality systems that aim at minimizing negative quality (such as poor service, inconsistency, mistakes). With those systems, the *best* you can get is *nothing wrong,* which we can see is not enough when all the players are good. Modern quality systems move beyond eliminating poor service to maximize positive quality (such as fun, luxury, comfort, ease of use). This creates *value.*

QFD is the only comprehensive quality system specifically aimed at satisfying the customer. It concentrates on maximizing customer satisfaction (positive quality) as measured by indicators such as return business, referred business and compliments. QFD focuses on delivering value by seeking out both spoken and unspoken needs, translating these into actionable services and communicating these throughout the organization. Further, QFD allows customers to prioritize their requirements, tells us how we are doing compared to our competitors and then directs us in optimizing those aspects of our service that will bring the greatest competitive advantage. What business can afford to waste its limited financial, time and human resources on services that customers do not want or on services where we are already the clear leader?

Types of Requirements

To satisfy customers, we must understand how meeting their requirements affects satisfaction. There are three types of customer requirements to consider (see Figure 10-1 and Appendix D, QFD and the Expanded Kano Model).

Revealed requirements (also called "spoken" performance needs) are typically what we get by just asking customers what they want. These requirements satisfy (or dissatisfy) in proportion to their presence (or absence) in the delivered service. Fast service would be a good example. If the customer is expecting fast service, the faster (or slower) the service, the more they like (or dislike) it. It is important that the customer (or market) segment you are trying to satisfy is truly understood. For example, a husband and wife going to an upscale restaurant in the evening to celebrate their wedding anniversary would probably

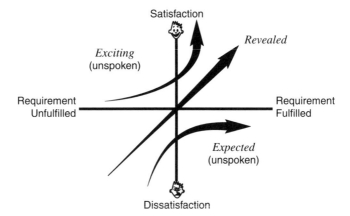

Figure 10-1 The Kano Model (adapted). World class services must meet all three types of requirements—not just what the customer says.

prefer to dine at a leisurely pace and would be upset by a pushy waiter trying to "turn" his tables as many times as possible in an evening. Given the same restaurant, the same man (or woman) and a different waiter at lunch time and there may not be the time or desire for a leisurely pace; in fact, the customer may be really upset if there is not prompt service allowing him or her to lunch and then get on with the afternoon's business.

Expected requirements (also called "unspoken, unless violated" basic needs) are often so basic the customer may fail to mention them, until we fail to deliver them. They are basic expectations of the service without which the service may cease to be of value; their absence is *very* dissatisfying. Further, meeting these requirements often goes unnoticed by most customers. For example, if an airplane takes off safely, passengers barely notice it. If it fails to take off safely, dissatisfaction, though brief, is intense. Expected requirements *must* be fulfilled.

Exciting requirements (also called "unknown" excitement needs) are difficult to discover. They are almost always beyond the customer's knowledge or expectations. Their absence does not dissatisfy, but their presence excites. For example, if champagne and caviar were served in coach class on a flight from Detroit to Cleveland, passengers would be ecstatic. If the eating fare were more mundane, passengers would hardly complain. These are the things that wow the customers and bring them back. Since customers are not apt to be aware of the potential for fulfilling these requirements, it is the responsibility of the service organization to explore customer problems and desires to determine opportunities for new levels of service.

Kano's Model is also dynamic in that what excites us today is expected tomorrow. That is, once introduced, an exciting service will soon be imitated by the competition and customers will come to expect it from everybody. An example would be special long-distance telephone rates at certain hours. On the other hand, expected requirements can become exciting after a real or potential failure. An example might be the passengers applauding a pilot who has safely maneuvered a landing despite severe weather conditions.

The Kano Model has an additional dimension regarding which customer segments the target market includes. For example, the champagne and caviar that might be exciting in the coach section might be expected on the Concord flight from New York to Paris. Knowing which customer segments you wish to serve is critical to understanding their requirements.

Thus, eliminating service problems can be likened to expected requirements. There is little satisfaction or competitive advantage when nothing goes wrong. Conversely, great value can be gained by discovering and delivering on exciting requirements ahead of the competition. QFD helps assure that expected requirements do not fall through the cracks and points out opportunities to build in excitement in the service offering.

THE KEYSTONE CUSTOMER

Many service organizations are part of a chain of customers. For example, an auto parts warehouse distributor purchases a muffler from a manufacturer and redistributes it to a retailer, who in turn sells it to a repair facility, which then installs it on a car driven by the customer's wife. The retailer, the installer and the customer are all part of a customer chain; they have different needs, occasionally conflicting ones.

QFD can accommodate multiple customers. The first step, though, is to uncover what is called the "keystone" customer (see Figure 10-2). Who ultimately determines the success or failure of our service? Like the keystone that holds a Roman arch in place, if we do not satisfy the keystone customer's needs first, the whole customer chain can collapse. In our muffler example, the keystone is the wife. If she is unhappy with the sound or smell of the car after the new muffler is installed, she may ask that it be checked again (time for which the installer will not be paid), and if she is still not satisfied, she may not want her car taken to that installer for other services. Conversely, if the keystone customer is satisfied, good will and word-of-mouth

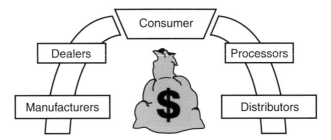

Figure 10-2   The Keystone Customer, who ultimately determines the success or failure of our service?

advertising may result. In QFD, it is important that the needs of the keystone customer be addressed first.

COHERENT SERVICE PLANNING

Once customer requirements are obtained, they must be translated into actionable plans and communicated throughout the service organization. This requires analyzing the customer needs for expected and exciting requirements, designing and planning new services and facilities, developing training programs and finally implementing the new service. Traditional development lacks the structure to communicate what matters most to the customer and to align organizational components and employees behind these critical requirements. Such a system is incoherent and inefficient. Thus, more time and resources are spent correcting and adjusting customer complaints than planning it right the first time (see Figure 10-3). This reduces profits for the service organization in two ways. In the short run the costs to operate (to get current service completed) are increased while customer satisfaction (and chances that future services will be sold to that customer) and revenues from service are reduced in the long run.

QFD IS COHERENT

When constrained by financial, time, human and other resources, when faced with regulatory, competitive and other pressures, it is necessary to concentrate the best efforts of all members of the organization on what matters most to the customer. It is necessary for these best efforts

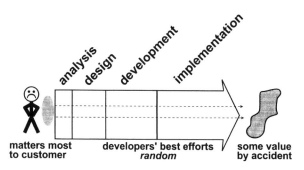

Figure 10-3 Incoherent Planning and Development. Traditional planning and development fails to focus best efforts. This is inherently inefficient, and dissatisfying.

to be aligned, or coherent. This way, each person builds on and reinforces the efforts of others to deliver what matters most to the customer (see Figure 10-4). The result is a superb service that exhibits features that have the greatest value to the customer.

To accomplish this, customer needs must be analyzed for unspoken requirements and prioritized. Then both the needs and the priorities must be translated into responses by the organization. The activities of each individual in the service organization are then developed accordingly, so that they may concentrate on the vital few aspects of their job *without constraint.* In effect, we "pull out all the stops" to satisfy our customers. This analysis, prioritization, translation and participation by everyone is called Quality Function Deployment.

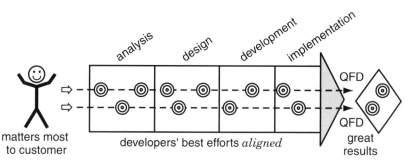

Figure 10-4 Coherent Planning and Development. QFD targets best efforts on value to customer. For equivalent effort, more value is received.

WHAT IS QFD?

Yoji Akao, the man who developed Quality Function Deployment with Katsuyo Ishihara of Matsushita Electric in 1965–1967, defines QFD as "a method for developing a design quality aimed at satisfying the consumer and then translating the consumer's demands into design targets and major quality assurance points to be used throughout the production stage." Change *production* to *service* and we might paraphrase this to "a system and procedures to aid the planning and development of services and assure that they will meet or exceed customer expectations."

QFD is a philosophy for quality assurance, not merely a series of steps to follow. The reduction of QFD to four phases in the West (as used by many practitioners) has prompted Akao, in the introduction of his latest book, to regret this "misapplication or incomplete use of QFD . . . that often elevates the mechanics of a product above customer satisfaction." Rather, a comprehensive QFD system "must reflect technology, reliability, and cost considerations" (see Figure 10-5).

The name QFD expresses its true purpose, which is satisfying customers *(quality)* by translating their needs into a design and assuring that all organizational units *(function)* work together to systematically break down their activities into finer and finer detail that can be quantified and controlled *(deployment)*.

THE TOOLS OF SERVICE QFD

While traditional quality tools were developed to handle quantitative data, new tools were created to handle the more qualitative language and relationships often associated with nonmanufacturing activities. The tools aid process re-engineering in improving existing services as well.

Matrix data analysis charts are used to present the results of multivariate analysis of data. Particularly for customer segmentation, techniques such as conjoint analysis, cluster analysis, factor analysis, multiple regression analysis and other techniques are useful when substantial quantitative customer data exist. This is the most mathematically sophisticated quality tool.

Affinity diagrams are used to surface the "deep structure" in voiced customer requirements. This rightbrained tool is generally pro-

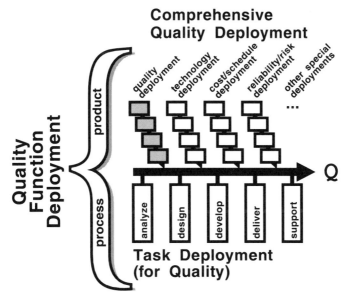

Figure 10-5 Comprehensive QFD is a System. QFD systematizes the improvement of quality, technology, cost and reliability of both the service itself and the process of planning and delivering it.

duced by the *KJ Method*™ developed by cultural anthropologist Jiro Kawakita (1986). Team members can directly elicit customers' natural organization of requirements. Also, this approach makes a good first step for creating hierarchy diagrams.

Relations diagrams (also called interrelationship digraphs) can be used to discover priorities, root causes of service process problems and unvoiced customer requirements.

Hierarchy diagrams (also called tree diagrams or systematic diagrams) are found throughout all QFD to check for missing data, to align levels of abstraction of the data, to diagram the why/how nature of functions and to diagram failures.

Matrices and *tables* are used to examine two or more dimensions in a deployment. Common types include relationships matrix, prioritization matrices and responsibility matrices.

Process Decision Program Diagrams (PDPCs) are used to analyze potential failures of new processes and services.

The *Analytic Hierarchy Process (AHP)* is used to prioritize a set of requirements and to select from among many alternatives to meet those requirements. This method employs pairwise comparisons on hierarchically organized elements to produce a very accurate set of priorities (Saaty, 1990; Tone and Manabe, 1990).

Blueprinting is a tool used to depict and analyze all the processes involved in providing a service (George and Gibson, 1991). It is a variant of the diagrams used in time/motion studies.

DEPLOYMENTS OF SERVICE QFD

Organization Deployment

This is used to map the QFD steps to the different organizational functions such as the president, Marketing and Planning, Development, Training, Customer Service, and so on. It shows who is responsible for what activities and when it occurs during the service planning and development process. Often, it is used with a responsibility matrix to clarify organizational roles (Chalmers, 1992; Mizuno and Akao, 1993; Nakui and Terninko, 1992b). This deployment is often ignored in the West, although, ironically, in Japan it proceeded the matrix deployments. It is highly recommended that Organization Deployment be done before QFD is applied to a specific service, so that the necessary team players understand their respective roles, activities and schedules. The tools used are flow charts and matrices. (see Figure 10-6).

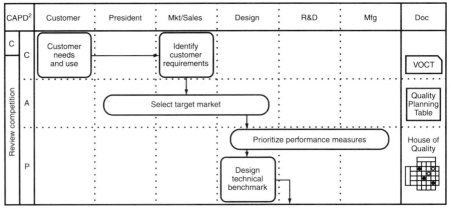

[1]Adapted from Nakui, S. and Terninko, J. "A Road Map to Better Design Process: Structuring a Quality Design Process Chart," *The Goal/QPC 9th Annual Conference: Advanced QFD Proceedings, 1992.*

[2]CAPD is a version of the PDCA cycle explained in Chapter 3

Figure 10-6 Organization Deployment. The Quality Design Process Chart defines organizational, responsibilities as well as determining which matrices, tables, etc. get done by whom and when.

Customer Deployment

This is the deployment of organizational goals (profit, utilization rate, etc.) into core competencies (skills, location, etc.) into customer attributes (high disposable income, impulse buyers, etc.) into target customer segments [yuppies, DINKS (Dual Income, No Kids), seniors, etc.]. This helps tailor our service offerings to the needs of those customers who can best help us achieve our goals. Unlike mass-produced products, services often focus on niche markets. The tools used are the AHP, matrices and matrix data analysis charts.

Voice of the Customer Deployment

These tables are used to record raw customer data, use characteristics, and separate the different types of service attributes, such as demanded quality, consistency, reliability, safety, and so on. These tables are also used to uncover unspoken customer needs such as expected and exciting requirements. The tools are VOC tables.

Quality Deployment

This is used to translate customer demanded quality and priorities into measurable service quality attributes such as accuracy, responsiveness, atmosphere, privacy, and so on. Targets can then be set for these attributes so that customer satisfaction can be assured. The tools are affinity diagrams, hierarchy trees, prioritization matrices, tables and the AHP.

Function Deployment

This is used to identify functional areas of the organization that are critical to performing tasks that must achieve the quality attribute targets. The tools used are affinity diagrams, hierarchy diagrams (function trees) and relationships matrices.

Process Deployment

This is used to diagram the current and reengineered processes. Blueprinting is the tool used.

New Concept Deployment

This is used in conjunction with Quality Improvement Stories (a structured problem-solving approach) to select a new process that will best satisfy customer needs (Hosotani, 1992; Imai, 1986; King, 1989a; Ozeki et al., 1990). The tools are blueprinting and concept selection matrices.

Task Deployment

This is used to break down critical jobs into tasks and steps. It identifies what the tasks and steps are, who does them, where they do them, when, how, how well (measurable standard), with what equipment, required training and skills and personality and human relations. The Task Deployment Table can be sorted to yield valuable information such as job descriptions, schedules, floor plans, standards, equipment and training requirements (Mazur, 1992; Mizuno and Akao, 1993). The tools are blueprinting and tables (e.g., a Task Deployment Table).

Reliability Deployment

This is used to identify and prevent failures of critical customer requirements. The tools are hierarchy diagrams (fault trees), PDPCs and relationships matrices.

CASE STUDY: JAPAN BUSINESS CONSULTANTS, LTD.

In 1982, while a part-time MBA student and a full-time automotive warehouse manager, the University of Michigan asked me to participate in a Joint U.S.–Japan Automotive Studies Project as a translator and interpreter. I did this on a part-time basis until my graduation in 1984. In 1985, I was selected for an interpreting job at the American Supplier Institute. My assignment was to work with a former Toyota quality specialist, Akira Fukuhara, while he taught an obscure subject called QFD. Over the ensuing three years, I spent several weeks each year interpreting for Mr. Fukuhara as he taught this methodology to the Big Three auto manufacturers and many of their key suppliers.

In 1986, I had the pleasure to finally meet the two men who created QFD, Mizuno and Akao. In 1987, Akao was invited by Bob King of GOAL/QPC to teach QFD in Massachusetts. Just *two weeks* before the start of a five-day seminar, King asked me to translate some QFD material and Akao began faxing over a series of twelve QFD articles that contained some of the most complicated charts I had ever seen. The charts covered a variety of industries from tractors to construction to software. Given the time constraints and a lack of technical knowledge of these varied industries, I knew my wife, Mayumi, and I could not do it alone. We needed more people who possessed language skills equal to ours.

Fortunately, I had been applying QFD to my business, and when this new opportunity arose, I had the tools to analyze the situation. By understanding King's needs, Akao's needs and, most importantly, the needs of GOAL's students, we were able to complete the translation of what eventually was republished as a 369-page book (Akao, 1990b). What follows is an update of this, the first application of QFD to a service in the United States.

CUSTOMER DEPLOYMENT

In 1984, I knew the translation business was what I wanted to do. But finding enough business for what was still a minor language was difficult. I needed to pursue customers that would bring success. The first step was to define success, which I will call organization goals. Since not all goals were of equal importance, it was necessary to prioritize them. I was not aware of the Analytic Hierarchy Process at that time so I had to guess. The results of my analysis are demonstrated in Figure 10-7.

The next step was to determine which of my skills (core competencies) could be best exploited to achieve my goals. A relationship matrix was set up using my goals and their priorities as inputs in the rows and my core competencies as outputs in the columns. The double circle,

	FI	Ex	Tim	Ln	Normalize Columns				Sum	%
Gain financial independence (FI)	1	2	7	7	0.56	0.63	0.37	0.4	1.96	**0.49**
Exploit areas of expertise (Ex)	1/2	1	9	9	0.28	0.31	0.47	0.51	1.57	**0.39**
Control my time (Tim)	1/7	1/9	1	1/2	0.08	0.03	0.05	0.03	0.19	**0.05**
Learn new knowledge. (Ln)	1/7	1/9	2	1	0.08	0.03	0.11	0.06	0.28	**0.07**
Totals	1.79	3.22	19	17.5	1	1	1	1	4.00	**1.00**

Figure 10-7 AHP of Organizational Goals. AHP uses pairwise comparisons and ratio scales to calculate priorities. [Saaty, 1990]

circle and triangle are used to symbolize the strength of the relationship, which is multiplied by the row weights. The resulting product is added for each column and normalized to a percentage at the bottom to yield which of my core competencies was most exploitable (see Figure 10-8).

Finally, I created a matrix between my now-prioritized core competencies and the types of customers I could pursue (see Figure 10-9). This led me to look for business with Japanese sales people coming to the United States to sell to the Big Three and with experts coming to teach Japanese quality and management. In fact, this is still the mainstay of our translation business eight years later.

WHATs vs. HOWs	Core Competencies	Japanese language (BA)	Business knowledge (MBA)	Automotive experience (10 yrs)	Service experience (7 yrs)	Teaching experience (3 yrs)	Access to Jpn experts	Priorities (AHP)
Strong Relationship: ● 9								
Medium Relationship: ○ 3								
Weak Relationship: △ 1								
Organization Goals								
Financial independence (Fortune)		●	○		●	○		49.5
Exploit expertise (Fame)		●		●		●	○	39.4
Control of time			●		○			4.6
Gain knowledge			○				●	6.5
Abs. Wt.		800	209	355	459	503	177	
Core Comp. Wt.		32	8	14	18	20	7	

(c) 1993 Glenn Mazur

Figure 10-8 Organizational Goals/Core Competencies Matrix. This matrix indicates that my Japanese language and teaching skills could be the most useful. Sounds like an interpeter's job.

WHATs vs. HOWs			Customer Segments	Automotive industry	Exporters	Management consultants	Translation agencies	Government	Japanese in U.S.	Core Comp. Wt.
Strong Relationship:	⬤	9								
Medium Relationship:	◯	3								
Weak Relationship:	△	1								
Core Competencies										
Japanese language (BA)				◯	◯		⬤		⬤	32
Business knowledge (MBA)					◯	⬤			◯	8
Automotive experience (10 yrs)				⬤	△		◯		◯	14
Service experience (7 yrs)							◯	◯	△	18
Teaching experience (3 yrs)					⬤				◯	20
Access to Jpn experts				⬤		⬤				7
Abs. Wt.				287	135	320	385	55	434	
Cust. Seg. Wt.				18	8	20	24	3	27	

(c) 1993 Glenn Mazur

Figure 10-9   Core Competencies/Customer Segments. Pursue Japanese coming to the U.S.

VOICE OF CUSTOMER DEPLOYMENT

Once my target customers were selected, the next step was to find out what they wanted. Having already worked with Fukuhara and Akao, it was evident they liked the United States. With the GOAL/QPC seminar, it was critical to get the translations done on time. These and other customer requirements needed to be analyzed using the Voice of the Customer Tables (see Figures 10-10*a* and *b*) (Nakui, 1991; Ohfuji et al., 1990). In VOCT Part 1, the voice "People want to hear what I have to say" is examined in terms of the use the client will have of my service. I discovered underlying requirements: "My message is accepted" and "I am asked to return often." In VOCT Part 2, the

Voice of Customer	Use	Reworded Data
People want to hear what I have to say.	Salesmen, teachers coming to do business.	My message is accepted. I am asked to return often.
Work done in 2 weeks.	Teaching class.	Work done in 2 weeks.

Figure 10-10a   Voice of Customer Table-Part 1 (partial)

Demanded Quality	Quality Attributes	Function/Task
My message is accepted.		
I build ongoing relationships. ←	Return often.	
My deadlines are met. ←	Done in 2 weeks. ←	Do work.

Figure 10-10b   Voice of Customer Table-Part 2 (partial)

reworded data or requirements are sorted by categories that will later be used to position them in the appropriate matrix deployment. If this were not done, then our matrices would be a jumble of different data and the resulting priorities would be misleading. The Voice of the Customer Tables were devised to avoid the problems of "misapplication or incomplete use of QFD," which Akao lamented about earlier.

When using VOCT Part 2, we should look for additional demanded quality items for use in our next deployment. Demanded quality items are the imprecise words that describe what it takes to satisfy the customer. Quality attributes are the measurable aspects of a service like frequency, turnaround time, and so on. When we encounter these, we should ask, "Why is this important to the customer?" Here, I found that my clients wanted to build ongoing *relationships* and that *deadlines* to be met.

All the demanded qualities were then grouped using the affinity diagram (see Figure 10-11). The hierarchy tree is not shown. The demanded qualities became the input rows to what is called the House of Quality Matrix, so named by Sawada of Toyota Auto Body for its many rooms and occasional roof (see Figure 10-12).

The rightmost room of the House of Quality is called the Quality Planning Table. It is here that customer priorities, competitive assess-

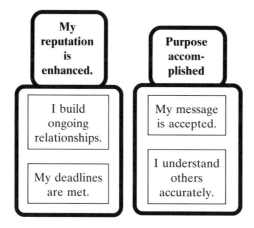

Figure 10-11 Affinity Diagram for Demanded Quality.

ments and company objectives are factored in to produce an overall weight. The Importance rating column is a scale of 1–5, with 5 being most important to the customer. Based on my discussions with clients, meeting deadlines was far and away the most important quality requirement for a translation service. Conventional wisdom might suggest that accuracy (no mistakes) would be most important. Sure enough, it is. In QFD, however, "no mistakes" is an expected requirement; we call this reliability, and as Figure 10-13 shows, it will have a deployment all to itself. Were we to include reliability and other expected requirements in the House of Quality, they would overpower all the satisfying and exciting requirements during the prioritization process. The result would be a service that had nothing wrong but not necessarily anything right.

The next three columns in the Quality Planning Table compare my existing service with my competitors. For "deadlines," given the GOAL/QPC Project I was facing, no one would be good enough. Still, since this was the most critical demanded quality, I chose to set a target of 5 (the best) on the same scale as before. Dividing where I want to be (5) by where I was (3), I calculated an Improvement ratio of 1.67. The Sales Point is another weighting factor that reflects the direction in which the organization wants to head. I chose a medium value of 1.2. The absolute weight was calculated by multiplying the Importance Rating by the Improvement Ratio by the Sales Point ($5 \times 1.67 \times 1.2 = 10.0$). The absolute weights for all the demanded qualities were summed and each one divided into the sum and multiplied by 100 to yield a percentage, which is called the Demanded Quality Weight.

This number represents the priority of each demanded quality based on Customer need, Competitive performance and the objectives of

WHATs vs. HOWs	Time	Turnaround time	Availability	On-time rate	Return visit rate	Repetition of Message	Expertise	Subject matter expertise	Social expertise	Language expertise	Importance Rating	Current Service	Competitor S	Competitor U	Target	Improvement Ratio	Sales Point	Absolute Wt.	Demanded Quality Wt.
My reputation is enhanced																			
I build ongoing relationships.			○	○	●				●	△	3	3	2	4	4	1.33	1.2	4.8	26
My deadlines are met.		●	●	●							5	3	2	3	5	1.67	1.2	10.0	53
Purpose accomplished																			
My message is accepted.					○	●		○	△	●	2	4	2	3	4	1.00	1.0	2.0	11
I understand others accurately.			○			○		●	△	●	2	3	2	3	3	1.00	1.0	2.0	11
Absolute Weight		479	587	555	262	128		128	251	217									
Quality Attribute Wt.		18	23	21	10	5		5	10	8									
Our Current Performance		10	10	90	2	3		3	75	10									
Competitor S		6	8	75	1	1		0	33	8									
Competitor U		9	8	90	5	2		3	50	10									
Target		20	24	100	7	3		3	75	11									
Unit of Measurement		Pg/day	Hrs/day	%	# Visits	# Repeats		# Courses	% Customs	FleschGr									

Relationship legend:
Strong Relationship: ● 9
Medium Relationship: ○ 3
Weak Relationship: △ 1

(c) 1993 Glenn Mazur

Figure 10-12   The House of Quality. The House of Quality consists of the Demanded Qualities, the Quality Planning Table, the Service Quality Attributes, the relationships matrix that transfers the Demanded Quality Weights into Service Quality Attribute Weights. At the bottom is a quantatitive comparison of these attributes for my company and two of my competitors.

the Company. I call these the three C's. These weights tell us not to concentrate on things that do not matter much to customers or where we are already ahead of the competition. The weights focus our improvement efforts on areas where we are not meeting customer needs and where our competition is better. This is one of the efficiencies of QFD; it causes you to concentrate your resources and efforts where they will have the greatest positive impact on the customer's demanded quality items.

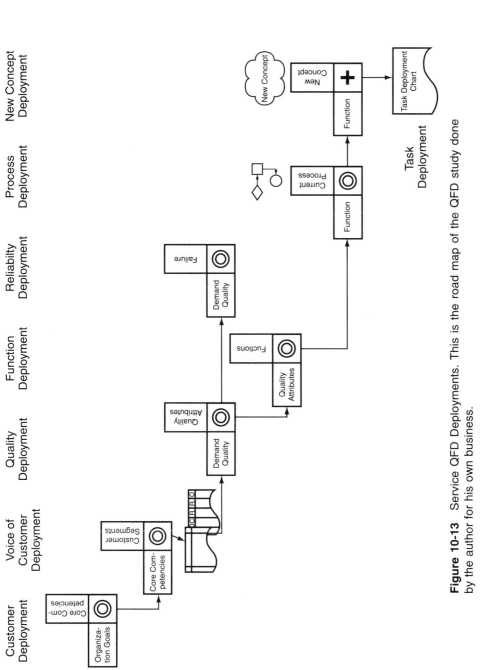

Figure 10-13 Service QFD Deployments. This is the road map of the QFD study done by the author for his own business.

157

The next step is to translate the demanded quality items into measurable service quality attributes. In QFD, assuring quality is a main objective, and unless something is measurable, it cannot be assured. Demanded quality items cannot be measured directly. For example, how would you measure "My message is accepted?" We have no gauge for acceptability. We must find a substitute that is measurable, Return Visit Rate, for example. We call these substitutes service quality attributes. Common ones are volume, convenience, responsiveness, efficiency, privacy, and so on (Brown et al., 1991).

The service quality attributes are also grouped with an affinity diagram and a hierarchy tree (not shown) and are entered in the columns of the House of Quality. I then examined each demanded quality to see which service quality attributes could help me measure it. If a relationship existed, I noted the strength of the relationship with a symbol: double circle for strong, circle for medium and triangle for weak. As in Figure 10-8, the demanded quality weights were multiplied by a value for the symbols (9, 3, 1), and the products (not shown due to software representation) were summed to give the absolute weight at the bottom of the House of Quality. These were then normalized to give the Quality Attribute Weights.

I then compared my measurable performance level for the most critical attributes with my competitors as best as I could measure based on what clients told me. The most critical areas dealt with the time I could be available to translate this mammoth project. Since the maximum time was two weeks, I set a daily target of working 24 hours a day. Right away, my wife pointed out a problem! I encountered what is called a technical bottleneck. To solve this bottleneck required further deployment and analysis.

FUNCTION DEPLOYMENT

I looked at the functions that I currently performed and that I would need to perform to satisfy my clients. I used a fishbone diagram (no subbones) with the demanded quality as the head and functions as the bones. For "My deadlines are met," functions such as Manage Communications and Output Data were determined. These were organized into a function tree. A second matrix was created with the quality attributes, their priorities and targets from the House of Quality as the input rows and the function tree as the output columns (see Figure 10-

Figure 10-14 Service Quality Attributes/Function Matrix

Legend — WHATs vs. HOWs:
Strong Relationship: ◎ = 9
Medium Relationship: O = 3
Weak Relationship: △ = 1

WHATs vs. HOWs	Manage Client	Manage travel	Manage communications	Communicate customer issues to client	Communicate client benefits to customer	Advocate client/subject	Solicit customers for client	Introduce subject matter to customer	Manage Information	Input data	Process data	Translate written information	Interpret spoken information	Output data	Study subject matter	Quality Characteristic Wt.	Target	Unit of Measurement
Time																		
Turnaround time												◎	◎	◎	△	18	20	Pg/day
Availability	O											◎	◎	◎		23	24	Hrs/day
On-time rate	◎		O						O				◎		◎	21	100	%
Return visit rate	△			◎		O						△	◎		O	10	7	# Visits
Repetition of message			O	◎		◎	◎						O		O	5	3	#Repeats
Expertise																		
Subject matter expertise	O					O							O	△	◎	5	3	#Courses
Social expertise	△		◎	△									◎			10	75	%Customs
Language expertise			O	O		O						◎	◎		◎	8	11	FleschGr
Absolute Wt.	279		205	169		134	114		432			660	474	357	182			
Function Wt.	9		7	6		4	4		14			22	16	12	6			
Function Targets	100% air & grnd		100% preview	100% explain		Call 1/week	Before meeting		Any type & time			n wrd/hr/person	2 people/client	Any format	3 hours			

(c) 1993 Glenn Mazur

Figure 10-14 Service Quality Attributes/Function Matrix. This matrix transfers Quality Attribute priorities and targets into Function priorities and targets.

14). Relationships were analyzed and the priorities transferred as in the House of Quality.

The most critical function was Translate Written Information and the quality attribute target of 24 hours per day was translated into a function target of n words per hour per person. I examined our current translating process of each person working on a project by themselves and realized that it was not possible to reach our target of n with this method. Before changing the process, I wanted to clarify possible failures so I could avoid them in the process re-engineering. I created a matrix with demanded quality items as input rows and potential failures as output columns (see Figure 10-15). This revealed that Late

WHATs vs. HOWs	Insulting	Inappropriate dress	Inappropriate language	Bad attitude	Late	Mistakes	Boring	Demanded Quality Wt.
Strong Relationship: ● 9 **Medium Relationship:** ○ 3 **Weak Relationship:** △ 1								
My reputation is enhanced								
I build ongoing relationships.			△	○		○	●	26
My deadlines are met.					●	○		53
Purpose accomplished								
My message is accepted.		△	●	○		●	○	11
I understand others accurately.			●	△		●		20
Absolute. Wt.	11	217	119	479	428	262		
Failure Mode Wt.	1	14	8	32	28	17		

(c) 1993 Glenn Mazur

Figure 10-15   Demanded Quality/Failure Modes Matrix.

and Mistakes were critical failures. This caused me to consider an extra edit step in any new process I might try.

NEW CONCEPT DEPLOYMENT

Keeping in mind my potential failure modes, I began experimenting with using multiple people on a team, new technology like transcribers, machine translation, remote translators linked by fax and modem, and so on. I created a New Concept Matrix with the key functions, priorities and targets as the row inputs and the new concepts as the column outputs, based on work done by Bob King and modeled after Stuart Pugh's system for generating new concepts (see Figure 10-16). Here, a plus means better than the current (or baseline) method, a minus means worse, and an S means the same. I then selected the best process that would perform the functions at their targets, was least costly and was least prone to errors. This was Concept 2, which used "new" technology: a transcriber so I could dictate the editing for a more natural sounding English.

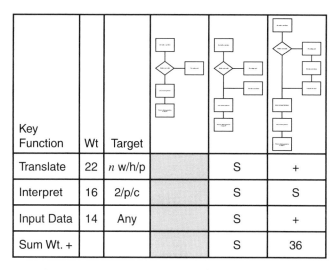

Key Function	Wt	Target			
Translate	22	n w/h/p		S	+
Interpret	16	2/p/c		S	S
Input Data	14	Any		S	+
Sum Wt. +				S	36

Figure 10-16 Key Function/New Concept Matrix.

What	Who	When	How	How Much	Why
Divide work	Glenn	At start of project		per # translators	to assure even flow and work given where translator competent with subject matter
Translate	Translators	14 hours/day		n word/hour	to assure completion in 2 weeks
Type	Typist	as available		150 wpm	to type handwritten work
Edit	Glenn	as typed	using transcriber	4 hours/day	to assure natural sounding English
Retype	typist	as transcribed	using transcriber	2 hours/day	to create final version
Send out	Mayumi	daily	fax	as available	so GOAL could do artwork

Figure 10-17 Task Deployment Chart.

What	Process Flow
Who	Job Description
When	Schedule
Where	Floor Plan
How	Equipment List Training/Skill Requirements Personality Requirements

Figure 10-18 Task Deployment Chart Sorted by Categories.

TASK DEPLOYMENT

The final step was to outline the necessary tasks, who would do what, when, where, how much, and so on. Figure 10-17 is a portion of that effort. Though not required here, the Task Deployment Chart can be sorted by its categories as follows (see Figure 10-18) to create other useful documents.

CONCLUSION

The job was done in two weeks. Akao was impressed because we actually completed the English work before the final Japanese monthly publication was printed. Japan Business Consultants has continued to grow in both revenues and the number of quality materials it handles. In fact, most of the source documents for Hoshin Kanri, TQM, Daily Management, QFD, Kansei Engineering and others have been translated by us via this process.

11

QFD AND
DESIGNING SOFTWARE

Software Quality Deployment (SQD) seeks to deploy the Voice of the Customer throughout the software engineering process so everyone involved hears the *Voice of the User*. Many of the problems of software engineering in practice can be traced to the difficulty of defining user requirements during the analysis phase of a project. Structured analysis techniques have been developed in an attempt to deal with this, but further improvement is needed. SQD strengthens the existing software engineering framework. This requires marrying quality deployment with contemporary software engineering techniques and professional project management practices.

Basic to application of software quality deployment is the use of a variety of matrices to examine in detail the interaction of various dimensions, such as function, cost, customer demands, facility structure, and so on. Quality deployment can be applied at various levels of sophistication, ranging from using just 4 basic matrices (Hauser and Clausing, 1988) to 30 matrices (King, 1987b) to over 150 matrices (Akao, 1988b). In this chapter on utilizing QFD for software design the focus will be on a basic set of matrices using the alphanumeric approach to identifying matrices (King, 1987b).

Although QFD does not have a long history of application in the field of software design, it has been used successfully for large and complex projects, such as the design of supertankers. It was this ability of QFD to assist in designing difficult and complex projects and to provide impressive results that caused it to be considered for use in one of the most complex, multifaceted product areas, that of software development. With some adaptations and enhancements QFD can be applied very effectively.

"VOICE OF THE USER"

What SQD does is to convey requirements from raw expressions given by the users to software engineering data and process models, and beyond, so that nothing is lost. From the most tentative expression of a want or need by a user to a delivered feature in an installed system, the user's voice must be heard and deployed. It is this deployment—a rigorous, visible communication of requirements—that defines and preserves the value or priority the user has for the proposed software capabilities. SQD starts with the source of the voice—the user.

User/User Requirements

QFD for hardware starts with "customer demands" (in the House of Quality or A1 matrix). For software, customer demands are the requirements of the users and other stakeholders in the project. Since software development projects must often serve several classes of users and stakeholders, it is necessary to clearly understand who has what requirements and to what extent. Users must first be identified, and their requirements determined and prioritized, before beginning work on the A1 matrix.

Identify Users. The potential users (and stakeholders) of the software are identified and organized into a hierarchy that reflects their interest in the project, not just any organizational relationship they happen to have. The Z0 matrix (see Figure 11-1a) may be used for this. There may be frequent and infrequent users or some that will not use the software at all but will determine whether it will be used. There may be internal and external users. In this step the "market segments" of users will be defined. This expanded notion of a "customer" covers users and stakeholders. For a system to be built and delivered, many people and groups must participate, review, approve and cooperate. All the wants, needs and concerns of these stakeholders must be understood to efficiently develop the best system for them, individually and collectively.

Determine User Requirements. The raw expressions of user and stakeholder wants, needs and concerns are gathered by interviews,

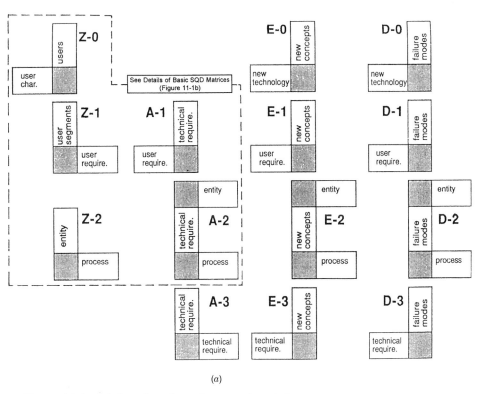

(a)

Figure 11-1 (a) A series of matrices used to define and deploy the "voice of the user" in SQD (b) Details of the basic SQD matrices.

surveys, Joint Application Design (JAD) or team analysis sessions, focus groups, trouble reports, problem logs and compliments on any existing systems. Analyzing current problems, potential opportunities and strategic directions of the organization are often fruitful sources of information. These expressions must be refined into clear, consistent statements of user expectations. Then using two of the seven management and planning tools (7-MP), affinity diagrams and relations diagrams (also called interrelationship digraphs) these expressions are organized into a final hierarchy of user requirements. This is accomplished with the Z1 matrix (see Figure 11-2). The relations diagram has proven especially useful in practice. A final hierarchy of only three levels such as created in this chapter has worked well even on very large software development projects.

Prioritize User Requirements. Once the user requirements hierarchy has been defined, the linkages from users to requirements can be made.

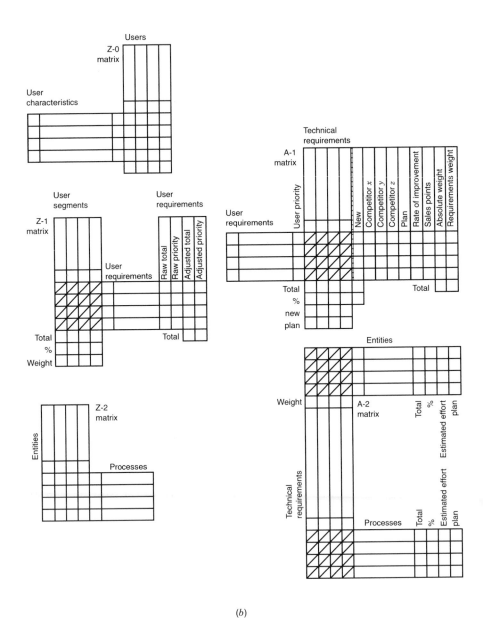

(*b*)

Figure 11-1 *(Continued)*

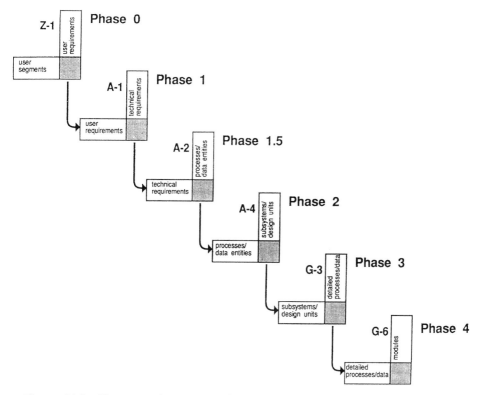

Figure 11-2 Phases and sequence of matrices used to deploy the "voice of the user" in SQD.

As part of this process, the user requirements are refined by quantification into more precise statements of expectation. Finally, the raw priority of all users is calculated. The Analytic Hierarchy Process (AHP) provides a robust and powerful approach for doing this with a scale of one to nine degrees of contribution (see Table 11-1). AHP was specifically developed for evaluating hierarchies. In use since the early 1970s in management science circles, AHP includes a consistency index and supports sensitivity analysis (Saaty, 1988).

Adjust User Requirements. As a practical consideration, it is often the case that the expectations of the users (and stakeholders) conflict or require more resources and time to address than are available. It may prove helpful to focus the project on the most valuable requirements by adjusting the raw priorities by an adjustment factor such as the number of users in each category (or their clout). This adjusted priority would

Table 11-1 Symbols Showing Full Scale of Degree of Contribution for SQD

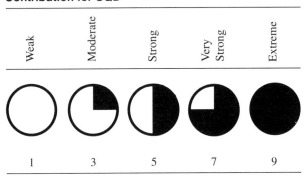

then be deployed to the A1 matrix. The raw priorities reflect what the users want most. The adjusted priorities reflect which users you want to satisfy most.

Knowing who the users and stakeholders are and what their wants, needs and concerns are is needed to understand why they have these needs, wants and concerns. To truly understand the Voice of the User, you must strive to understand the mind of the user. "Why do they want this?" must be asked repeatedly. Only then can the user requirements be accurately and precisely stated. Mere listening is not enough. Understanding the problems, opportunities, priorities and value is necessary. The reason that the Voice of the User must be reviewed again and again is that the user will often state that a certain feature is necessary in an application. He or she will say this because past experience indicates that without that particular feature there will be certain problems or there will be difficulty accomplishing some task. It is necessary to question the users to the point that you understand the underlying problem and not accept as necessary some solution that addresses the underlying problem.

User Requirements/Technical Requirement

With the users' expectations accurately reflected in a prioritized hierarchy, they must be translated into technical requirements. This is the job of the House of Quality or A1 matrix (see Figure 11-3).

Determine Technical Requirements. The processing and data necessary to support the user requirements must be determined. As software can only process data, the "substitute quality characteristics" must state

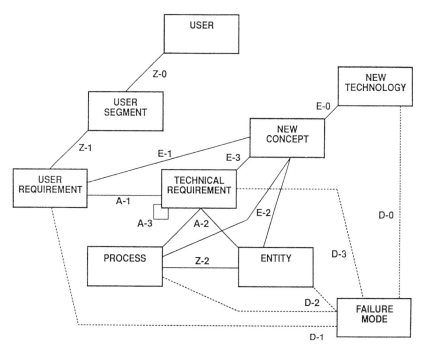

Figure 11-3   Entity/Relationship Diagram, used to define data requirements of SQD.

what can be done to contribute to a user requirement (and only in terms of data and processing). Once these technical requirements have been identified and organized into a hierarchy, the linkages from user requirements to technical requirements can be made. The AHP 1–9 scale works well, and it is easier to use a single scale throughout the entire SQD process. During this step, the technical requirements are refined by quantification into more precise statements of the data and processing required.

Adjust User Requirements. For software the user priorities may require further adjustment to accurately reflect the competitive position of the software and the selling points necessary for a user to "buy" the software. The competitive position adjustment takes into account the performance of this software versus other software available to the user. The level of performance now, what is appropriate for a target or plan and the rate of improvement necessary to compete in the future are also considered. The sales point adjustment highlights what is required to attract users to the software. These may be features such as packaging, which may persuade users to buy (or try) the software but may not affect their satisfaction in using the software (Snodgrass and Kasi, 1986). After adjustment to reflect competition and sales points, the final user requirement weights will be deployed to all subsequent matrices.

Prioritize Technical Requirements. Once the user requirement weights have been calculated, linkages from user requirements to technical requirements can be calculated. The resulting total and percent (calculated by dividing the individual score by the total weight for each technical requirement) for the technical requirements are the final priority. The current level of technical requirement performance now and a target or plan level can be stated quantitatively. In this way the user priorities are accurately reflected on the technical requirements that get implemented.

Now the relative priority of each technical requirement—the data and processing the users need—is known. The next step is to deploy the technical requirements to software engineering models for analyzing the processes and data entities in detail.

Technical Requirements/Processes/Entities

Modern software engineering practice requires a thorough and detailed modeling of the technical requirements. This requires mapping the technical requirements to data models and process models, represented by some type of Entity–Relationship Diagram (ERD) and Data Flow Diagram (DFD). The A2 matrix handles this in a T-matrix style (see Figure 11-1*a*).

Determine Entities. The data component of the technical requirements are separated and analyzed in detail on an ERD. The data entities and relationships required to support the technical requirements are identified and their contribution indicated using the AHP 1–9 scale. As part of this step, the entities are refined into more precise statements of *what* data are required but not *how* to implement them.

Determine Processes. The process components of the technical requirements are also separated and analyzed in detail on data flow diagrams. The processes required to support the technical requirements are identified, and their contribution is indicated using the same AHP 1–9 scale. Similarly, the processes are refined into more precise statements of *what* processing is required but not *how* to do it.

It is now possible to determine the value to the users of a particular process or entity. For each process or entity, the contributions to each technical requirement are known and visible. The contribution of each technical requirement to each user requirement is also known and visible. Thus, we can precisely see the value of a particular process or entity from both a competitive perspective and a user/stakeholder perspective.

Processes/Entities

The process model represented on a data flow diagram and the data model represented on an entity/relationship diagram must be consistent with each other. To ensure this, the Z2 matrix maps entities against processes (see Figure 11-1a). Every process must receive its input data and have a destination for its output data, which calls for specific data in specific entities. Similarly, the data in every entity must be created, retrieved, updated and deleted by specific processes. This matrix can be done while the diagrams are being created and refined. The Z2 matrix is an essential cross-reference between the process and data models.

Other matrices in the format of the A2 matrix may be useful depending on the analysis approach for your project. An object/entity/process matrix would support an object-oriented analysis approach. An event/entity/process matrix would support an event-oriented analysis approach. A state-and-transition/entity/process matrix would support a real-time analysis approach. The entity and process information in matrix A2 links to detail on the actual models. For a sparse matrix, a table format would work just as well. A table for the purpose of integrating data and process models is common in current software engineering projects.

SQD is a user-centered approach to software development. It concentrates on maximizing user satisfaction from the software engineering process. The key to applying QFD to software is to have a solid understanding of the modern software engineering models and techniques. What SQD does is to integrate them so nothing is lost, from the most tentative expression of a want by a user to the installation and maintenance of the software. The tools and techniques of software engineering and project management are the foundations of SQD. Software can only be as good as the process that produces it (Zultner, 1988). SQD is one way to directly enhance the software engineering process.

SOFTWARE ENGINEERING

Developing software by engineering is still a young engineering discipline and is not practiced as widely as it might be (Malsbury, 1987).

Despite this, a number of rigorous software engineering techniques are well established for requirements definition. These structured analysis techniques allow precise definition of analysis requirements and design specifications in a visible way that can be inspected, planned and managed. This is an essential prerequisite for fully applying QFD to software with the most effective results.

Software Engineering Models

In analyzing the user requirements for a system, there are three basic kinds of requirements to define: process, purpose and data. There are three standard (or static) models used to capture these requirements (see Figure 11-4).

Process. To define process requirements, DFDs are used (see Figure 11-4). These diagrams are logical (not physical) in that they show only what is required, not how to accomplish it. Three levels of diagrams are necessary (and are usually sufficient). A high-level (level-0) DFD documents the activities of major importance (within the scope of the project). A midlevel (level-1) DFD documents the normal and routine activities. Low-level (level-2) DFDs specify all the detailed tasks, errors

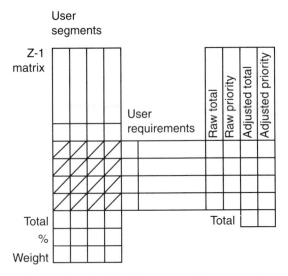

Figure 11-4 The Z1 matrix, showing the source of user requirements: users and stakeholders.

and exceptions. These lowest level processes are supported by rigorous process specifications in one of a number of formats: a decision tree, a decision table, structured English (which may be in the form of an action diagram), tight English (which may be in the form of a script) or an equation (Gane and Sarson, 1979).

Purpose. To define the purpose of the project, objectives are formally stated. The interdependencies of objectives can be graphically shown on an Objective–Hierarchy Diagram (OHD) (see Figure 11-4). The supporting objective descriptions are carefully stated in precise narrative form—"formal English" (Zultner, 1989).

Data. To define data requirements, an ERD (see Figures 11-1c and 11-4) is used. This documents the entities (significant "things" about which we wish to keep information) of the organization (within the scope of the project), how they are related (the relationships) and the associated data attributes (also called data elements, i.e., individual "pieces" of data). Each entity, relationship or attribute is precisely defined in a supporting data dictionary, which also defines all the data shown on the DFDs (Howe, 1983).

In analyzing the user requirements for many systems, there are other kinds of requirements that may be present, such as control, events and data access/usage requirements. There are supplemental (or active) models used to capture these requirements (see Figure 11-4).

Control. To define control requirements, Control Flow Diagrams (CFDs) are used. These diagrams show the control of the processes (on the DFD) and explain the logic of that control. Control processes are supported by control specifications, usually in the form of tables, which depict the actions to be done under all valid conditions (Hatley and Pirbhai, 1987).

Events. A very powerful way to look at systems is to view them as a planned-purpose mechanism responding to stimuli. The stimuli and response constitute an event. The events a system must handle, and what it should do in each case, are usually stated in an event list or event table (see Figure 11-4) (McMenamin and Palmer, 1984).

Access. When large amounts of data are used and accessed, especially if this occurs frequently or quickly, it becomes necessary to define usage or data access requirements. This is usually done with an Access Diagram (AD) (see Figure 11-4), for which a wide variety of notations exist. This documents the specific requirements for access to data in terms of inputs, outputs, frequencies, volumes, priorities, and so on. The data dictionary is generally used to support the diagram with detailed information (Schuldt, 1986).

These six basic software engineering models provide the means for the user requirements (that SQD captures, organizes, refines and prioritizes) to be deployed throughout the software engineering life-cycle. Thus, SQD acts as a strong front-end to further integrate the well-established models of modern software engineering with the user at the center (Zachman, 1986).

Software Engineering Enhanced

As users typically cannot articulate all their requirements, it is the responsibility of the software engineer to "ask why five times" to define and analyze the users' requirements at a fundamental level (Ishikawa, 1986). We must ask why they do what they do, why they currently have the problems they do and why haven't they seized their opportunities. Only by getting close to the users and understanding their wants, needs and concerns can a complete set of requirements be confidently prepared. To satisfy users, we must understand how meeting their requirements affects satisfaction (Kano et al., 1984b). There are three types of user requirements to attend to (see Figure 11-5): expected, normal and exciting requirements.

Expected Requirements. These *musts* are those features the user expects: Their presence meets their expectations but does not satisfy them. Their absence, however, is very dissatisfying. Failure to meet an expected requirement may make meeting all other requirements inconsequential. On-line help is an example for some users. Since users assume it, they may not think to mention it when they are asked about their wants, needs and concerns. They may even be unaware a system could possibly be delivered without it.

Normal Requirements. These *wants* are satisfied (or dissatisfied) in proportion to the presence (or absence) of the feature. Swiftness of operation is an example for some users. The faster (or slower) the system is, the better they like (or dislike) it.

Exciting Requirements. These *wows* are features beyond the users' expectations and are highly satisfying when present and well executed. Their absence does not dissatisfy because they are not expected. These

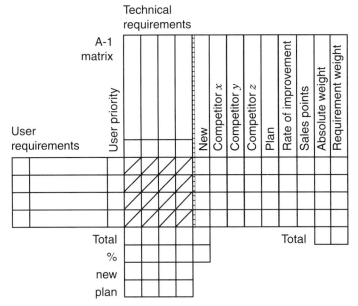

Figure 11-5 The House of Quality A1 matrix, deploys the WHYs of user requirements into the WHAT of technical requirements.

are often key selling points or means of differentiating the product. For some users a window-and-icon environment is an example. Users will show off exciting requirements to their peers.

Traditionally, software engineering efforts have concentrated on essentially minimizing user dissatisfaction from the software development process. The focus was on detecting defects (by inspection, testing or logging complaints), looking upstream to analyze the causes of the defects and then working to eliminate those causes from the software engineering process. This approach seeks to minimize users' complaints (negative quality). The best that can be done with any defect detection approach is zero defects. That is not good enough, because the absence of negatives (defects) only makes a system less bad, that is, not dissatisfying (Deming, 1986).

Today, software engineering using SQD concentrates more on maximizing user satisfaction from the software engineering process. The focus is on preventing the causes of defects through a deeper understanding of the user's true requirements starting with a careful study of user and stakeholder wants, needs and concerns. With SQD, you work upstream to design quality into the system and continuously work to improve the software engineering process with innovation. This approach seeks to maximize the users' compliments (positive quality). Only strong positives can make software so good that users boast about it—the true test of exciting quality.

In working with several organizations in applying SQD experience to date indicates that a solid foundation is required in software engineering. This provides the software-specific engineering know-how and project management that provide the planning and coordinating of time and resources necessary to build good software (see Figure 11-6). SQD relates the previously disparate models of software engineering into a meaningful framework. By providing a means to trace any technical detail or requirement back to raw expressions of the user, the source of the need is clarified, as is the site of downstream use. When technical trade-offs are made deep in design, the designers can trace the relevant technical issues back to the user's needs and take them into account when making technical decisions. This helps to prevent user dissatisfaction later, when the system is delivered. Everyone involved knows there are users, listens to what they want and tries to understand why they want what they want.

QUALITY FUNCTION DEPLOYMENT

Many organizations using QFD for their nonsoftware products and projects are experiencing significant results. Yet applying QFD to software is not as straightforward as it might at first appear. The adaptations and enhancements required for software may also prove interesting and useful for your future nonsoftware projects. In addition, SQD can be a powerful avenue for improvement intervention by software/quality professionals in the software engineering area.

QFD Adaptations

Software development has some significant differences from hardware and manufactured products. This requires some adaptations to QFD for software use. The factors of material and material-based costs are replaced by factors of data and time (or schedule). Some structural changes are also required. The result is a revised QFD framework suitable for software (see Figure 11-1*b*).

Data Replaces Material. In QFD a number of matrices deal with material. Software development is perhaps unique among engineering areas in that no raw material is necessary to produce a product. Software

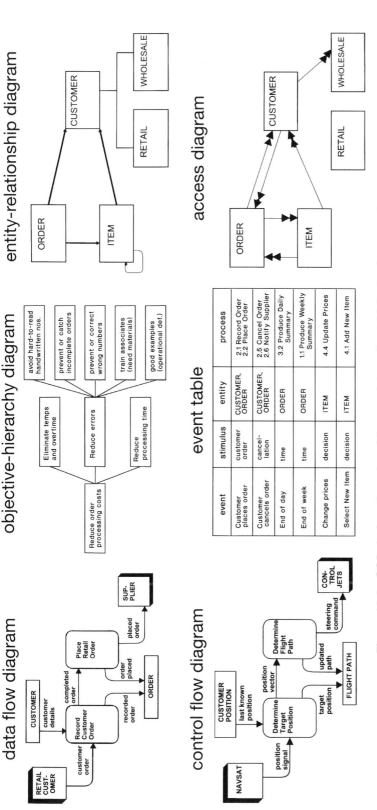

Figure 11-6 SQD deploys the voice of the user into well-established software engineering models.

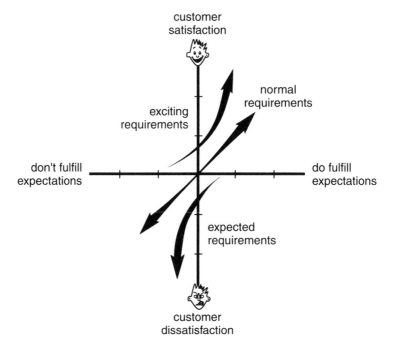

Figure 11-7   User requirements must capture all three types of requirements for completeness (Kano Model).

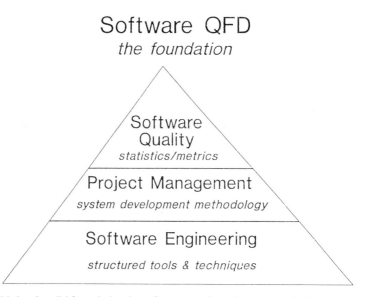

Figure 11-8   A solid foundation in software engineering and project management is necessary for full benefits from SQD.

can be produced *directly* from requirements—a pure example of the deployment of requirements into a product. The nearest analog to material for software is data, and a data dictionary is not unlike a data inventory or data bill-of-materials. Thus data replace material in SQD matrices. The one difference is that data are coequal with process in importance and thus occur very early. For nonsoftware situations the material matrices tend to occur later in the QFD process.

Time Replaces Cost. With no raw material, the primary cost of software is labor. So time determines cost, which is why the project management focus during software development is on time. Substituting time for cost in the appropriate matrices works nicely for software, if the data are available for the length of time it takes (at a detailed level) to do each software development task. The B-series of matrices are used for schedule reduction efforts, but these are beyond the scope of this chapter.

Process Replaces Function. The most difficult aspect of applying QFD to software is understanding how the function =→ mechanism =→ part (or product =→ systems =→ subsystem =→ component =→ part =→ raw material) decomposition works for software. Since software is not material—only instructions to a computer—there is not just one obvious way to decompose functionality. There are very few physical constraints on software, so almost *any* decomposition can work. As this decision is crucial for the A1 matrix, how shall we proceed? As a practical (and very effective) measure, the high/mid/low levels of requirements (see DFDs mentioned earlier) work extremely well. In fact, since the levels are defined in terms of user importance, it may be easier for users to understand and work the initial matrices for SQD than for nonsoftware QFD. It also makes it easier to focus on the significant/important requirements earlier and pursue them (alone) in greater detail. If you think of the user requirements as a tree of progressively finer detail (and a rapidly increasing number of branches), the tree can be pruned in-process (more so than for many nonsoftware applications) in SQD.

What and Why. The House of Quality, or A-1 matrix, is usually explained as *what* versus *how*. This is premature for software. The WHATs must be understood in more detail with structured analysis. The WHATs must be decomposed into process and data requirements and each analyzed with additional techniques. Process requirements are analyzed with a process model—DFDs. Data requirements are analyzed with a data model—ERDs. Both models are *logical* models; they are free of any particular design or implementation approach (including whether a computer will be used). Thus for software, the A1 matrix must be WHY versus WHAT. Considering HOW in an A1 matrix is too early for software. Software design, where the HOWs are decided, occurs much later.

No Roof. With the A1 matrix a WHY-versus-WHAT matrix, a roof comparing WHAT versus WHAT is far less useful than in nonsoftware QFD. As we will not consider how the technical requirements might be done until later during design, it is difficult to find any technical requirements that conflict in an A1 matrix. At this point, the WHATs can only conflict conceptually, which is the only implementation free conflict possible. Is this of value to analyze? The answer is generally not. It may be needed for a particular project, and then the A3 matrix is used.

Data/Process Model Deployment. Simply listing the WHATs, even organized in a hierarchy, is not adequate for software. We must understand the interconnections of the processes as well as must understand the relationships of the entities. So the technical requirements must be deployed into the software engineering models for process and data. The A2 matrix does this. Since there are two models to be deployed, a T-matrix style is used instead of the more common L-matrix style of hardware QFD.

Determining Relationships. In filling in the relation of WHYs to WHATs in the A1 matrix, the degree of contribution must be expressed. The use of traditional Japanese symbols (O O Δ X #) is widespread for hardware QFD. But these symbols have no meaning for most Americans. It is desirable that the symbols used be intuitively understood and provide for a full range of evaluations. Numbers (1, 3, and 9) can be used but do not offer a visual "response surface" of the degree of contribution. One example of a full-range symbol set that has worked well in practice is shown in Table 11-1.

New Concepts. Technology is a powerful force for innovation and exceeding user expectations. The D0 matrix (see Figure 11-1a) supports the careful and visible consideration of new technology for new concepts. The E-series of matrices (see Figure 11-1a) allows consideration of new concepts for features and uses of the software. This has great value during design when design alternatives are explored (see Figure 11-1a).

Failure Modes. Reliability is an important concern as software becomes more and more an essential support tool for organizations. The D-series of matrices examines the potential failure modes and how they can be prevented or ameliorated. It is especially important to carefully examine the possible ways new technology can fail with the E0 matrix. Together the D0 and E0 matrices allow for a balanced understanding of new technology, the benefits and risks of design alternatives (see Figure 11-1a).

Adapting QFD to software is a continuing effort. The work of Bob King provides an excellent foundation (King, 1987b). Beyond the basic SQD matrices presented in this chapter, the B-series have great potential for software project management, and the C-series are also useful for software design. But SQD requires not only adapting QFD but enhancing it. Some of these enhancements may prove useful for non-software QFD applications.

QFD Enhancements

Some QFD tools and techniques have been refined or enhanced for SQD (see Table 11-2). A few are briefly described below.

Data Flow Diagrams. Flow charts were widely used in software circles 20 years ago. For about the last 10 years, they have been supplanted for analysis purposes by DFDs. These can be powerful tools for non-software applications. There is nothing "technical" about them. They are used in software engineering to analyze activity in an organization.

Table 11-2   Tools of QFD

Existing Tools	Improved Tools
Flowchart	Data flow diagram
Histogram	Stem-and-leaf plot, or digiplot
Tree diagram	Hierarchy diagram
Arrow diagram	Precedence diagram (time scaled)

Precedence Diagrams. In Japan the arrow diagram (also called activity-on-arrow diagram) is taught as one of the 7-MP tools that accompany QFD (Mizuno, 1988). The arrow diagram was the first project management diagram developed for PERT in the late 1950s. Later, the activity-on-node diagram was developed for CPM. Today the (time-scaled) precedence diagram is preferred by project management professionals and is readily available from PC-based software packages (Moder et al., 1983).

Analytic Hierarchy Process. A key aspect of QFD is prioritizing customer (user) demands. But asking people to rank isolated items along an arbitrary scale is a process vulnerable to many biases. A more robust approach is the Analytic Hierarchy Process (AHP), which uses pairwise evaluation by hierarchy level. It works for the rankings and the weightings and provides a measure of judgment consistency as well as sensitivity analysis (Saaty, 1988). AHP can be used during the creation of the hierarchies as an aid in pruning them to the most vital items and to a manageable size. There are many places to apply AHP in SQD; prioritizing failure modes is just one example. Here too software support is available in a product called Expert Choice.

Software Support. There is software to support most of the individual tasks of SQD but not the entire process. Existing Computer-Aided Software Engineering (CASE) tools support the software engineering aspects reasonably well but do not yet address the quality deployment side. As this is a highly competitive market, this should be expected to change in the next few years.

QFD enhancements for SQD are a good example of continuous improvement applied to the tools, techniques and process of improvement. Just as in decades past the Japanese did not merely copy American quality approaches, Americans should not merely copy, but understand and improve, the state-of-the-art Japanese tools, techniques and approaches to QFD.

Software Quality Assurance

SQD can provide Software Quality Assurance professionals with a way to focus improvement activities on two key areas: the products of systems development and the processes of developing systems.

Product Improvement. In order to know where to concentrate walk-throughs and other appraisal techniques during analysis and design, it is necessary to know what processes and entities provide the greatest value to the user (Freedman and Weinberg, 1982). Similarly, testing can be focused on those modules and data segments of greatest value to the user. Acceptance testing also should focus on the most valuable functions and data (Myers, 1979). The most valuable parts of a system in the eyes of the user should be the most trustworthy.

Process Improvement. Software development has some significant similarities to manufacturing. In manufacturing, it is acknowledged that improving the manufacturing process simply by buying better machines or tools is not enough. The distinctive characteristic of world-class manufacturing organizations is not their state-of-the-art tools—those anyone can buy. The critical factor in their success is the way they manage what they have, whether by Just-in-Time, total quality control or drum-buffer-rope principles (theory of constraints). Their *process* is better (Schonberger, 1986). To deliver world-class software requires a world-class software engineering process. This calls for a thorough understanding of *why* each step in developing a system is done and what *value* it adds for the user. Find the value drivers in your system development life cycle, improve them and you have strengthened the software engineering value chain for all software projects (Rappaport, 1986). Additional SQD matrices (the F-series and G-series; see Figure 11-1*b*) provide a means to identify and make visible the value drivers and value chain of software development.

Walk-throughs, testing and the software engineering process all benefit from visibly deployed, accurate user requirements. As users demand world-class software, software engineers must learn to listen more closely to the voice of the user and strive to understand the mind of the user. One effective way for Software Quality Assurance groups to lead this effort is by example, by applying SQD to their own projects and users.

The first project done with SQD usually takes more effort (about 15–20%) due to the learning curve. The quality and effectiveness of the training are key factors in minimizing this "investment." By the second or third SQD project, a net reduction in effort should be apparent (15–40%). The most important result is not the size of the improvement, significant though it is, but that it is a *sustainable* improvement. The knowledge captured by the SQD process is rigorously defined and visible: It can be leveraged and reused. The biggest beneficiaries appear to be the less-experienced software developers who can, by using the SQD matrices from similar projects, effectively operate with a much higher level of "virtual" experience. Software maintenance professionals can apply SQD even faster since their "cycle time" is much shorter.

Using an adapted and enhanced version of QFD, SQD can capture in a visible form the wants, needs and concerns of users and stakeholders and is a powerful approach for delivering better software faster.

12

QFD AND BUSINESS PROCESS RE-ENGINEERING

The QFD structure and methodology have traditionally been used when designing or redesigning a product or service. QFD is also very useful when designing a process or a group of interactive processes (a system). When designing a product or service, the initial matrix helps to organize the Voice of the Customer (the WHATs) and show how it will be translated into the technical requirements (the HOWs) necessary to address the WHATs. In later matrices, the design of the process(es) that will be used to deliver the product or service is addressed.

In the case of Business Process Re-engineering (BPR), however, it is the *initial* matrix that addresses the design of the process (because the process *is* the product). In addition, the use of QFD is especially important because the QFD matrices visually show what the new process design must accomplish to be successful in the eyes of its customers (the "ends") plus they provide detail on the techniques that will be used to address the customers' requirements (the "means").

Using QFD reduces the risk of failure of BPR while increasing BPR's effectiveness. Risk is reduced because there is a greater and continuing focus on the needs of the customers (both internal and external) of the process being re-engineered. In addition, there is much greater involvement by the organization's personnel (whose inputs are sought and deployed), especially on the part of internal customers. The inputs needed for doing QFD are established at the outset of the BPR engagement and are agreed to by the various customer sets. This early activity establishes the scope of the BPR project and is important because "scope creep" is a major cause of BPR project failures. Effectiveness is increased because the QFD matrices focus the project on doing the right things.

One of the most frequent reasons cited for failure of all types of change programs is the lack of communication and understanding between (a) the persons who will be impacted by the changes and (b) the group involved in creating the new process(es) and associated changes. This is particularly true of BPR, where the changes are large and discontinuous ("quantum change") and have been formulated and implemented over longer time frames with different groups of internal and external people involved. To a lesser extent, the communication problem also applies to Continuous Process Improvement (CPI) initiatives ("incremental change"). It is mitigated with CPI, however, because the changes are small with short time periods from formulation to implementation with both stages done by the same group. The impacted persons and the group involved in creating the process improvements are, at least, subsets of each other so there is less chance for poor communication, lack of information or misinformation.

Although the perception is that CPI efforts go on continuously whereas BPR projects are one-time events, the fact is that both CPI and BPR should be part of an organization's strategy for continuous change and adaptation to (exploitation of?) the ebb and flow of the marketplace. The use of QFD to support the first BPR project in a particular operational area allows for much greater definition and control, lower risks, and so forth, but a major additional advantage to using QFD is that later, when re-engineering is initiated a second time, it can be accomplished much faster and easier, thus gaining competitive advantage for the firm.

Studies on BPR have shown that the greatest gains come when core processes are reengineered on a companywide basis (or even go beyond the company bounds, i.e., partnering with suppliers and/or customers). Hall, Rosenthal, and Wade (1993) indicate that the results reported from doing BPR vary from lackluster to revolutionary. [Gadd and Oakland (1996) state that only one in three BPR efforts is successful. Other authors state failure rates of 50–75% (Caron, Jarvenpaa, and Stoddard, 1994; Hammer and Champy, 1993).] In many instances BPR efforts are successful when examined taking the narrow view of the resulting new processes or process improvements; that is, the order entry cycle time is reduced by 50% or first-time yield is increased by 30%. However, when the organization is viewed as a whole, there isn't much to brag about! Hall, Rosenthal, and Wade (1993) state that for an organization doing BPR to actually show major changes in the bottom line, four critical issues must be addressed.

Two are largely behavioral; the other two are mostly analytical. All four must be defined and integrated in the successful BPR project:

Behavioral

(a) *Strong leadership* from senior executives who are actually involved in the BPR efforts, setting high-level goals and then monitoring the development process

(b) *Communication* efforts throughout the organization explaining the reasons or drivers for change and how the change is being designed in response to those drivers

Analytical

(c) *Breadth,* defined in a way that addresses performance improvement on a companywide basis through increased customer perceived value and/or reduced cost

(d) *Depth* of redesign, which must reach the company's core, changing six key organizational element: roles and responsibilities, measurements and incentives, organizational structure, information technology, shared values and skills

The strength of QFD is in its ability to aid the definition and integration of these four critical issues. Let us take the issues listed and discuss how QFD can assist BPR.

(a) *Strong leadership.* QFD can take the goals of senior executives and use them as inputs (they are internal customers and stakeholders). The development of new processes (and the system) related to the BPR project can then be monitored as the re-engineering team progresses through the series of matrices. The matrices show what has been accomplished and what still needs to be done and shows the links between the different elements (see "Depth" below) of the project.

(b) *Communication.* QFD methodology uses the wants, needs and requirements from an organization's customers to create new systems and processes. The matrices explicitly show progression from customer inputs (the WANTs) to technical answers (the HOWs and HOW MUCHes) and all of the elements of the technical answers. By picking the appropriate level and focus, the matrices can be used to show

1. To lower level personnel how they will fit into the new system and why the processes they will work with when everything is changed are designed the way they are

2. To senior executives how an overview of the processes of the new system are integrated (as well as the reasons for that integration)
3. To middle managers and supervisors how they can highlight their own set of matrices showing linkage between high-level goals and low-level activities

(c) *Breadth.* The QFD matrices can be set up to represent all of the processes in the system and have specific matrices for addressing high-level cost and customer value issues for the system as a whole or decomposed for the individual processes that make up the whole.

(d) *Depth.* The use of matrices to describe the inner workings of the six key organizational elements *and* the linkages (feedback and feedforward) between them can help a BPR design team organize the area of greatest complexity. These relationships must be clearly understood and shown if software is to be part of the BPR solution.

Chris White's (1996) article presents a different focus on CPI versus BPR. He discusses the relative simplicity of a CPI team improving a task or a cell in a process or a part of a process. He then compares this to the complexity of BPR improvement efforts associated with, not just improving a process, but potentially changing it entirely, even eliminating or linking it to another process in a new way, for example, parallel versus sequential processing. To understand the leverage that QFD gives the BPR effort, you should recognize the level of complexity associated with redesign of a companywide core process (or a system composed of several processes) versus improvement of a task or cell. The first is about the relationships between processes (and even system-to-system links) and the other is about the small tasks, steps or operations that make up a process.

A complete BPR project consists of the following steps (see Figure 12-1):

1. Select a core process or system that has subnormal performance or offers competitive advantage in the marketplace; then discuss the issues around defining the scope of the process or system (starting point, ending point, for example).
2. Identify key customer groups of the process or system; establish the scope of the BPR project with the help of both internal and external customers.
3. Capture the customer requirements (the WHATs) for each group using different approaches (surveys, focus groups, interviews) to

① Select Core or Differentiating Process
② Identify Key Customer Groups of the Process
③ Capture Customer Req'ts. for each Group: WANTs
④ Link Customer Req'ts. to Process' Outputs

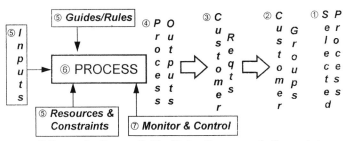

⑤ Determine Inputs, Guides/Rules, Resources & Constraints
⑥ Define Process: HOWs
⑦ Define Means to Monitor & Control Process

Figure 12-1 Business process re-engineering steps.

gather information on the different kinds of quality (expected, normal and exciting).

4. Translate the customer requirements into the requirements for process outputs.

5. Determine (a) the relevant guides/rules/regulations (industry standards, processing guides, company rules, government regulations) as well as (b) the available resources (personnel, money, time, facilities) and constraints (could be limited number of personnel with certain skills or capacity limits on capital equipment or just a minimum time to cure adhesive or dry paint or grow bacteria.

6. Define the process elements (the HOWs) [or processes (system)] that can consistently deliver a quality output in a timely manner, given the guides/rules, and the available resources and constraints.

7. Detail the necessary inputs needed to support the process.

8. Create a means to monitor and control the process(es) and/or system and then maintain the initial performance level as well as foster continuous improvement.

Just as analysis tools such as SPC applied in manufacturing processes contribute to the rise of CPI (also known as Total Quality Management, Continuous Improvement Initiative, etc.), the development of information technology made the impact of BPR possible. The application of tools like these have allowed (even encouraged) personnel directly involved in an operation to rapidly and easily access the exact information necessary for them to make decisions, and this has led to their empowerment. In addition, modern information technology has enabled BPR gains because it allows substituting business and technical information (which is cheap and can be made rapidly available throughout the organization in many useful formats) for expensive assets or constrained resources.

For example, greater use of marketing and sales forecast information in combination with early demand data gathering and analysis in a manufacturing environment can lead to lower inventory levels with higher service levels. This results in greater customer satisfaction, lower carrying costs and fewer charge-offs due to obsolete or out-of-season stock. In administrative processes and in the service sector, well-designed information systems allow a process with many steps to be compressed into one or two steps with one person able to handle several tasks simultaneously. This is due to "expert systems" assistance and instantaneous access to large, structured information bases. This approach results in dramatically reduced cycle times while increasing the quality of both (a) the work done and (b) the customer interface (and related customer satisfaction).

Although information technology is a major part of most BPR projects in some way or another, and because it offers tremendous possibilities for quantum improvement, it is difficult to work with successfully. To take time to define a system (complete with software for gathering, analyzing and distributing information) and then build it from scratch no longer makes sense because of the many flexible software packages on the market today. However, because of this quantity of software packages, with each offering a variety of different sets of features and functions, it is difficult to determine which is best for a particular information technology application and the system outcomes we are seeking without using some decision-making structure. The information in QFD matrices can help build the requirements for the BPR team to use in establishing the review criteria for standard software packages. Also, QFD will assist in establishing the extent to which there are gaps between the standard package and the needs of the system. Information from some of the QFD matrices used in the BPR can be used as inputs to software design matrices (see Chapter 11).

The difficulty with leveraging the application of information technology is that the flow, augmentation and decomposition of information from any but the smallest, simplest system is difficult to conceptualize, design and implement. Using QFD assists in defining what the relationships should be, defining the entire system and then measuring and monitoring the system as it is implemented. QFD can be of major assistance in a BPR project because of its organizational and analytical structure.

13

SOME UNIQUE APPLICATIONS OF QFD

QFD practitioners come in a variety of flavors: from marketing, engineering, manufacturing, assembly, quality and material and, much to the surprise of some, from finance, facilities and human resources. Because of the diversity of backgrounds of QFD practitioners, some rather unique applications of QFD have been recorded.

This chapter is provided as an inspiration to its readers, experienced and new QFD practitioners, who may believe that the application of QFD is limited to the design, development and production of tangible products or intangible services or processes. The authors have experienced some unique cases where QFD has been extremely successful in resolving unusual challenges.

We have found that QFD can easily be used in conjunction with both the analysis and creation of organizational missions statements. In either of those two situations the WHATs are designated as organizational commitments to its customers and the HOWs are the organizational actions/responses to meet the needs of its customers. Customer importance values and organizational risk values are added. Then the interrelationships between organizational actions/responses and the organizational commitments are determined. Finally, a combination algorithm is applied (sum the products of each customer importance value and the organizational risk value and their interrelationship value and then rank order the sums). The highest ranking organizational actions/responses are, or should be, a significant part of the mission statement along with the organizational commitments.

In the world of education and training, QFD has been used to create an entire curriculum, for example, an undergraduate major in a specific discipline or an MBA program. Even the contents of a specific course have been identified using QFD to prioritize specific topics from a long list of potential topics.

The modification of an existing software package as well as the creation of a totally new one has been enhanced using the QFD process. Creation of an individual's job description is well within the capability of QFD. One such application required the use of a four-matrix combination, as shown in Figures 13-1 through 13-4.

When an organization gets to the point where its corporate headquarters has created an excessive quantity of policies and procedures, it is time to zero-base the collection. This can be done using QFD to designate which are today's existing customer requirements and which policies and procedures relate to the requirements. Proceeding through the usual QFD steps, what remains is a prioritized listing of those policies and procedures that should remain on the books and those that should be canceled.

Another unusual QFD application focuses on situational analysis. The following outline and the four cascaded matrices in Figures 13-1 to 13-4 tell the story:

		Customer Importance	Continuous Training and Education			Constancy of Objectives	Management Visibility Without Fear		ORC/QSC Credibility	
							Company President Visibility			Sharing Information
			Team-work Skills	Leader-ship Training	Quality Respons-ibility Awareness	Common Priority/ Commit-ment	Active Involvement in Projects	Listening	Commun-ication Forum (Feedback)
Deming Philosophy Implementation	Organizational Development	5	5	5	5	5	5	5	5
Clear Vision		5	3	1	1	5	3	5	5
Management Issues	Deming vs. Hammer / Dept. vs. Deming	3	5	2	1	3	5	5	3
Communications		5		1		1	5	5	5
Absolute Weights			55	44	33	54	80	90	84
Percent			12	10	7	14	18	20	19
Key Elements						X	X	X	X

Figure 13-1 Leadership.

Customer Importance		Concise Written Communications	Action Item Awareness	Common Goals	Memo Format	Breakdown Interdepartment Barriers	Monthly Project Feedback	Total Quality Visibility	Visibility of SPC Success	Creative Communication Forum	Company Progress Review	Proper Preparation and Planning	Management Visibility	Better Understanding of Other Functional Areas	Physical Location of Offices	Better Listening
Better Interdepartmental Communication	9	3	9	9	3	9	3	1	1	3	3	3	3	9	3	9
Improved Communications from Upper Management	9	3	3	9	1	3	3	9	1	3	9	1	9	3	1	9
More Effective Meetings	3	1	3	1	1	3	1	1	1	1	1	9	1	3	1	3
More Effective Written Communication	3	9	1	1	9	1	1	1	1	1	1	3	1	1	1	1
Absolute Weights		84	120	188	55	120	60	96	24	60	114	72	114	120	42	192
Percent		6	8	12	5	8	4	7	2	4	8	5	8	8	3	13
Key Elements			X	X		X		X			X		X	X		X

Figure 13-2   Communication.

Customer Importance		Analyze Environmental Monitoring Data							
		Chart Envir. Monitoring Test Results and Post In Areas	Apply SPC to Environmental Monitoring	Evalute Need for Sampling	FDA Require-ments Commit-ments to FDA	Environ-mental Education	Managers to Review Charts	Facility Modi-cation (Up-grades)	Identify Critical Monitor-ing Areas
Identify/Remove Special Causes of Variation	5	5	5			3			
Understand Environ-mental Responsibility	3				1	5	3		
Corrective Action Follow Up	3	3	3		1	3	3	3	
Reduce Amount of Sampling	5	1	5	5	5			3	5
Facility Requirements Planning	5			1	3	5		5	5
Discussions/Proposals with/to FDA	5	3	5	5	5			3	3
Absolute Weights		54	34	55	71	64	18	49	65
Percent		12	18	12	15	14	4	11	14
Key Elements		X	X	X	X	X		X	X

Figure 13-3   Process capability improvement: environmental monitoring.

Customer Importance		Certificate of Analysis (COA)	Improve Consistency of Release (Planning) Eliminate Hot List	Testing		Work With Suppliers Work with Suppliers	Work with Suppliers to	Specs Better Understand
		Speed of Supplier COA Program	Minimize Release Time	Eliminate Duplicate Testing	Status Release Raw Materials	Work with Suppliers to Reduce Variability	Work with Suppliers to Determine Critical Characteristics	Better Understand Specs vs. Product Quality
Vendor Certification	5	5	1	5		5	5	
Realistic Release Test Requirements	5		5	5		3	5	5
Improve Release of Materials On Hold (Disposition)	3		1	1				5
Absolute Weights		25	33	53		40	50	40
Percent		10	14	22		17	21	16
Key Elements				X		X	X	X

Figure 13-4 Process capability improvement-customer/supplier relations.

Situational Analysis Using QFD

1. Brainstorming necessitated by
   a. Problem buildup
   b. Recognized need for continuous improvement
2. Brainstorming session included
   a. First-line supervisors
   b. Production and process (manufacturing) engineers
   c. Middle management
      (1) Environmental monitoring
      (2) Customer/supplier relations
3. Brainstorming followed by affinity analysis
4. Affinity analysis surfaced specific major management issues
   a. Leadership
   b. Communication
   c. Process capability improvement
      (1) Environmental monitoring
      (2) Customer/supplier relations

Another unusual QFD application, the Improvement Planning Table (Figures 13-5 and 13-6), was developed by Dan Neumann, Director of Consulting at Organizational Dynamics, Inc. (ODI). It is used to pin-

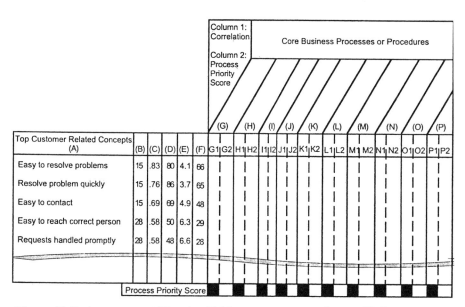

Figure 13-5 Improvement Planning Table: customer care abouts. (A) *Top customer care abouts:* selected, based on priority score (column F) from the list of all items form all transactions (PR = problem resolution, IR = information requests, B = billing). (B) *Percentage affected last 12 months:* percent of customers. (c) *Importance:* based on multiple regression analysis involving all items from all transactions. (D). *Percent not highly satisfied:* percentage of respondents answering less than 8 on a 0–10 scale, where 0 means "strongly disagree" and 10 means "strongly agree". (E) *Customer mean score* (F) *Priority index:* computation, C × D. (G–O) to be completed by management team, these processes/practices/policies are those most closely correlated to top employee care abouts. (G1–O1) *Relationship Score:* between the process and care about (1 = very low, 5 = medium, 10 = high); to be completed by management team. (G2–O2) *Relationship score:* process priority score: computation, F × G1 or F × H2 or F × I1, etc. (P) Sum of scores in each column G2, H2, I2, etc.

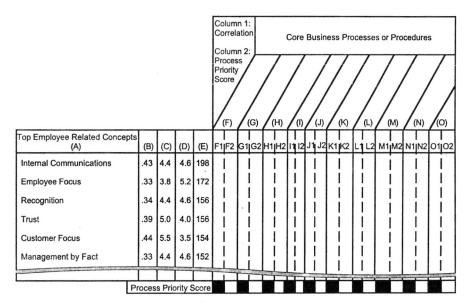

Figure 13-6 Improvement Planning Table: employee care abouts. (A) Top 20 employee concepts. (B)*Importance:* based on multiple regression analysis involving all concepts. (C)*Employee score:* employee's concept score. (D): *Mean gap.* computation, 9.0 − concept score. (E): *Priority index:* computation, B × C × 100. (F−N) To be completed by management team, these processes/practices/policies are those most closely correlated to top employee care abouts. (F1−N1): Correlation between the process and care about (1=very low, 5=medium, 10=high); to be completed by management team. (F2− N2): *Process Priority Score.* Computation, E × F1 or E × G1 or E × H1, etc. (O) Sum of scores in each column, F2, G2, H2, etc.

point core business processes or procedures within an organization that impact both customer- and employee-related concerns.

The tables in Figures 13-5 and 13-6 show the essence of this process. The top customer- and employee-related concerns were obtained from a well-designed and executed internal and external surveying process. The importance scores, mean scores and priority index were calculated by the computer program used to analyze the survey data. These top customer- or employee-related concerns form the left side of their respective matrices.

A list of core business processes or procedures was developed and prioritized and then the corresponding relationship scores developed. Once these relationship scores were completed, they were multiplied by the priority index and then summed by each column. The core business processes or procedures were then rank ordered. Those core processes or procedures were then analyzed by the management team, which sets improvement goals and commissions teams to develop action plans to achieve the stated improvements.

A major manufacturer of automotive circuit boards developed their Just-in-Time (JIT) manufacturing philosophy using QFD and fed the results into their Policy Deployment process. QFD was used to translate high-level business goals (related to JIT) into tangible action plans to be implemented throughout the organization. Corporate headquarters decreed that all plants develop and begin implementing a JIT plan. The circuit board manufacturing facility decided to use QFD and Policy Deployment approaches to create their JIT plan. A top-level management team was formed. The team identified and prioritized Critical Success Factors (CSFs). These CSFs were plugged into the QFD matrix as the WHATs. The CSFs were benchmarked and a five-year improvement goal was established. To know where to focus its attention, the team then correlated 18 JIT initiatives (i.e., the HOWs) to the prioritized CSFs.

The output from the QFD matrix identified which JIT initiatives were most important to the greatest number of CSFs and to the CSFs ranked highest in importance. That information fed the plan for deployment as well as set targets for the various JIT initiatives. Then an operations level planning meeting was held where the initiatives were displayed as input to a matrix. For each of the selected initiatives, projects that would support that initiative were brainstormed. These projects were then prioritized based upon how they impacted the various JIT initiatives (and the underlying CSFs). Then the relationships between the various projects were reviewed. In some instances, although a particular project was not high priority, it was preparation for or led to a more critical project, so it was placed on the Policy Deployment schedule anyway.

In another unique application, an aerospace contractor rationalized the design of a specific component to be used on each version of its entire product line. This component was absolutely necessary for every program in the contractor's product line. It served the same basic function in each program and was exposed to similar variations in temperature and thermal and mechanical shock and vibration. There the similarities ended. Because every program had engineered its own components, they were produced in very small quantities. In addition, there was very little ability to share reliability information and evolve toward a more reliable component. One contractor, with the goals of reducing costs and increasing reliability, proposed designing a common component for use on all of their systems. Operational and research information was gathered from all the contractor's program offices. From these user inputs a series of QFD matrices were constructed that resulted in a common component design that is now available for use in conjunction with all follow-on and new program designs.

III

APPLYING QFD ON AN ONGOING BASIS

14

APPLYING QFD
TO ROBUST
QUALITY SYSTEMS

A truly value-added application of QFD is in the field of quality systems analysis, design and implementation. QFD identifies interdependencies and provides a systemic perspective of Total Quality Management (TQM) and ISO 9000 models (ISO, 1987a) and the interrelationships between the ISO 9000 model elements. Using this QFD-based approach, an organization can define, clarify and exploit the synergies available between multiple philosophies, different quality award criteria and various customer and other requirements on the one hand and the stable, globally recognized quality system framework of the ISO 9000 models on the other. The approach is intended to be flexible and allow for adaptation to a wide variety of organizations as well as provide for modification and expansion as the organization grows and needs change.

As the economy becomes more global and increasingly competitive, new initiatives and approaches to increasing productivity and quality while reducing costs and cycle times command the attention of executives, managers and quality professionals. These personnel and, to a lesser extent, the supervisory and floor operators have had to contend with all sorts of concepts, ideas, programs and requirements, all bent on improving operations in some way. In confronting these multiple inputs and requirements, the often exasperated or frustrated managers view each acronym or new initiative as an additional layer of complication and documentation and are often stymied by inconsistencies or contradictions. Often these new requirements or initiatives are addressed by creation of goal-specific, independent teams (e.g., TQM,

Kaizen, poka-yoke, value engineering, re-engineering, Baldrige, ISO 9000 and QS-9000).

In some organizations, these teams may actually spend most of their efforts competing for limited resources and management's favor while working at cross purposes with other similar groups. As an aside, this may be one source of the discontent associated with many TQM or other improvement initiatives. Frustrated and harried managers fail to see and utilize the commonalties or synergies in developing and implementing their plans, further confusing or complicating their systems and inefficiently or ineffectively using their resources. This occurs even without the creation of multiple teams.

A case in point involves the intent of companies to implement a locally defined robust quality system, often entitled Total Quality Management (TQM) or Continuous Process Improvement or an organizationally identified variant system (e.g., Galactic Widgets Inc. Super Quality System) while also addressing and incorporating the required elements of an ISO 9000 series model standard. For purposes of discussion, the desired robust quality system shall be referred to as TQM.

Total Quality Management drives the understanding that a truly dynamic organization needs to continually redefine its customers' wants, needs and expectations, both known and unknown, and to marshal the limited resources of the organization to address these customer imperatives with maximum impact. With this is the understanding or awareness that to become more effective and productive in an increasingly competitive environment the organization must look to its own internal processes. The activities associated with TQM are based on leveraging a constantly growing knowledge base, but they still demand a fairly stable structure and a system that guards against overcorrecting. The organization must ensure that its structure, policies and procedures and activities are focused and coordinated. In a difficult business environment with ever higher expectations for return and profit, these organizations are containing costs, downsizing, streamlining or otherwise reducing staffs. These actions limit or constrain the resources needed to meet these increasing external and internal demands.

In addition to an already difficult situation, there comes the real or perceived need to address the requirements contained in the ISO 9000 series of quality system standards. This may be driven by a concern over exports, particularly to the European Union (EU). The EU has legislation, regulations and directives that specify or option ISO 9000 requirements in design, production or both through the modular approach to regulated product directives and regulations compliance. A growing number of domestic business sectors are establishing their own individually created variants of ISO 9000 models as "common" requirements between companies within their sector. The most visible of these is the automotive sector's "QS-9000" standard.

These concerns may be informed and actual, perceived or misinformed. The need may also be driven by contractual or other requirements from current or potential customers both overseas and domestic. For most companies, the need is individually defined by product, process, customer base and business strategy. Unfortunately, not every company fully defines these criteria strategically but blindly pursues the "herd instinct" or builds barriers to change.

In approaching all these demands, the usual response is to see them as an addition to and sometimes separate from existing systems. The perception prevails that they require additional resources of time, money, personnel or even new systems. Systems are cast and recast, with the resulting turmoil in operations reflected as the next perceived, or regularly occurring, "program of the month." The symptoms of this are the myopic concern with documentation, the development of program-specific "quality manuals," the aforementioned special focus teams and the resort to canned computer programs, manuals, procedures, data bases and templates that do not reflect or support the vision, goals or culture of the organization. These chaotic actions are mistakenly viewed and embraced as "implementation." Often, there are resultant complaints against an "innocent" ISO 9000 model when, in fact, it is management, especially project management and task analysis, that is at fault.

For the most part, all this confusion and chaotic change is unnecessary and often wasteful, if not harmful. Most well-conceived, functioning quality systems *do not* need to be overhauled extensively with each new set of external requirements. They do not need to be dressed up or even disguised by new garb consisting of renumbered, reformatted or repackaged documentation. Any quality system must be just that, a "system," a "whole" that is not just a loose accumulation or conglomeration of add-ons. A strategic, critical review and evaluation of the existing quality system in light of any new or revised requirements is needed to focus resources only on those areas that merit attention. This review and examination must also include all other internal or external requirements to reduce complexity and realize the synergies that will become apparent when the system is viewed holistically.

An approach, methodology or technology is needed to assist in creating this holistic view. Fortunately, one is available and already widely practiced. This approach is the set of tools and techniques associated with QFD. With this approach, a series of relational matrices cascade or deploy requirements through policies, functions, activities and documents in a top-down, progressive fashion. (See Figures 14-1 and 14-2.) Most QFD applications take product or service requirements or market factors and use the information to develop features, characteristics and/or specifications. These applications then deploy these features, characteristics, and so forth through engineering, processing, manufacturing, service and delivery functions of an organization. The focus is on the customer-purchased deliverable, whether it is a manufactured product or a service.

However, instead of a designing a product or service, QFD can just as easily be used on a design effort for the quality system itself. In place of the customer- or market-driven requirements (the WHATs), a set of TQM outputs such as those identified by the ASQC Quality Management Division's TQM Committee are compared to the 20 ISO 9000 series elements (the HOWs). Application and practice have shown that it is best to start by mapping the TQM outputs to the major paragraphs and items in the overview and guideline documents: ISO 9000-1, ISO 9004-1, and ISO 9004-2 (ISO, 1987a,e). Include each of the relevant TQM outputs (from ASQC): Quality Orientation, Continuous Improvement, Satisfy Customer Requirement, Long Term Mission, Management Led Improvement, Defect Prevention, People as Assets, Teamwork, Quality Partner Suppliers, Cross-Functional Team Efforts, Employee Empowerment, Concurrent Engineering, Data-Driven Decisions and Management by Planning. This ensures the consideration of the broader systemwide elements and the more descriptive or so-called generic aspects of the quality system.

A subsequent matrix maps either the TQM outputs, the ISO 9000-1/9004-1/9004-2 elements or both to the more specific directions contained in the ISO 9001, 9002, or 9003 (ISO, 1987b–d) standards or models. The matrix already contained in ISO 9000-1 can be used as a basis for this step, with the stringency noted in the matrix substituted for the common terminology of "strong," "medium" and "weak" used in traditional QFD applications. Experience has also shown that the ISO elements need to be augmented so as to achieve all desired TQM outputs.

At this time, it is also helpful to expand this matrix or create another that links the TQM outputs and/or the ISO elements with any regulatory or customer requirements. Industry-specific (e.g., the automotive sector QS-9000) or customer-specific interpretations, additions and modifications to the basic ISO models must also be recognized and captured

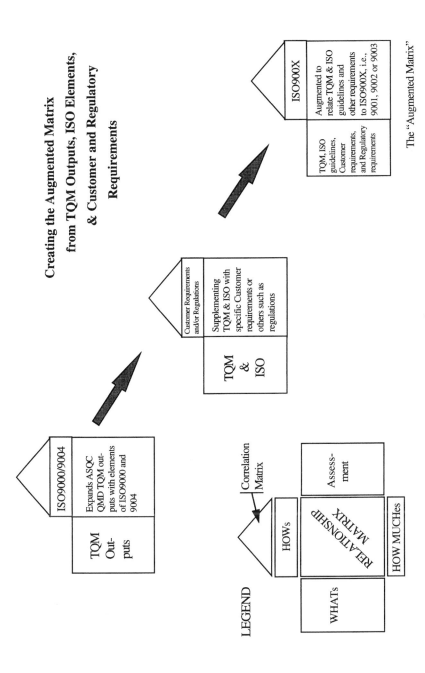

Figure 14-1 Creating the augmented matrix for TQM outputs, ISO elements and customer and regulatory requirements.

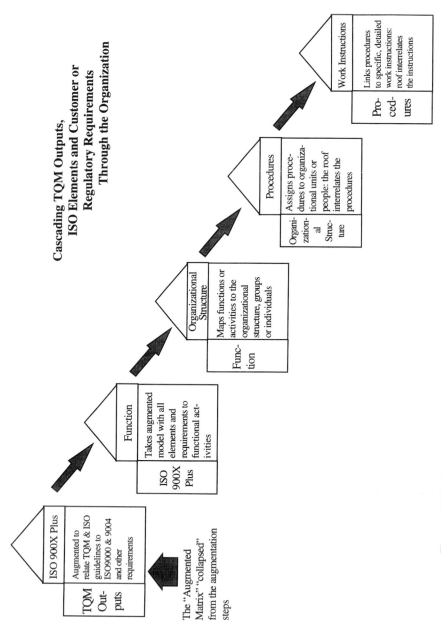

**Cascading TQM Outputs,
ISO Elements and Customer or
Regulatory Requirements
Through the Organization**

ISO 900X Plus — Augmented to relate TQM & ISO guidelines to ISO9000 & 9004 and other requirements

TQM Outputs — The "Augmented Matrix" "collapsed" from the augmentation steps

Function — ISO 900X Plus — Takes augmented model with all elements and requirements to functional activities

Organizational Structure — Function — Maps functions or activities to the organizational structure, groups or individuals

Procedures — Organizational Structure — Assigns procedures to organizational units or people: the roof interrelates the procedures

Work Instructions — Procedures — Links procedures to specific, detailed work instructions: roof interrelates the instructions

Figure 14-2 Cascading TQM outputs, ISO elements and customer or regulatory requirements through the organization.

in this QFD quality system model. Once this has been accomplished, the similarities, synergies and contrasts can be examined among the various requirements. The main body of the matrix relates the TQM outputs to the various additional applicable standards or requirements. The top, the "roof" of the House of Quality, can be used to correlate the various standards or requirements. For convenience, this step may be accomplished with a set of smaller submatrices, one for each element: design, marketing, process control, inspection, calibration, and the like. However, caution must be exercised to ensure that submatrices are not viewed as independent entities. After they are individually reviewed, they must be recombined for review of their interactions and adequacy as a system. Failure of organizations to benefit from implementing an ISO model frequently arises from their perspective that the ISO elements are independent activities, whereas they are actually interdependent. For example, the extent and detail of procedures and work instructions (elements 4.5 and 4.9) are dependent on the user's previous knowledge, education and skills (elements 4.1, 4.2, 4.6 and 4.18). The user's education, knowledge, linguistic skills and native tongue also define the appropriate textual content of documents. By taking the time do this analysis, the various requirements are seen as a whole interrelated quality system. Omissions, conflicts, inconsistencies, similarities, and so on can be identified, examined and resolved. This puts the entire effort into perspective and identifies and avoids potentially wasteful allocations of resources. It also forms the basis for discussions with various customers about their quality requirements to reach some resolution of conflicts, different prescriptive approaches to obtain the same results or non-value-added requirements.

In addition to the ISO 9000 focused matrix, which relates the elements of ISO 9004 to each of ISO 9001, 9002 or 9003 models, there are other useful cross-references and matrices. These relate elements of one or more of the ISO models or standards to other industry-specific standards. Care should be taken when completing a relationship matrix comparing an ISO model or output of the ISO-focused matrix and other standards. The ISO standards are unique in that they are "descriptive" and "generic," describing basic features or elements of a quality system. Essentially, ISO models define the WHATs. Industry-specific standards typically are much more "prescriptive." Often, they detail the HOWs, the HOW MUCHes and the HOW OFTEN types of considerations. Coding the relationships among the various standards and requirements should make important distinctions. There are cases where similar wording describes essentially the same basic requirements. However, in other cases, the wording merely indicates a particular, specific quantitative measure, frequency or level of performance and not an entirely new consideration.

Using the output of the augmented matrix linking TQM, ISO and other customer-specific standards, a subsequent matrix deploys the requirements to functional areas, such as general management, accounting, marketing, sales, human resources, engineering, manufacturing, maintenance, materials management, shipping, and so on. This matrix, which is customized to reflect *your organization,* details the interrelationships and dependencies among the organization's functional areas deploying the TQM or ISO elements. Examination of either the main interrelationship matrix or the correlations among the functional areas in the roof reveals omissions, duplications and potential conflicts. The main body of the matrix shows those functional areas with primary and secondary responsibility for each TQM or ISO element.

From this, another matrix can be developed linking the functional areas to specific activities. Those specific activities that are key to a particular functional area should be identified differently from those in which the functional area merely plays a supportive role. Omissions, duplications or potential conflicts are apparent at this point. As with the other matrices, activities having no prime functional area (i.e., no "home") or functional areas with no primary activities can be identified. These activities and functional areas become natural targets for examination and potential consolidation or elimination through re-engineering or restructuring. Needless to say, the documentation associated with these activities and functional areas must be examined and revised accordingly. Also, the roof expresses the relationships among activities and shows correlations between them. Activities that support and enhance each other can be compared to those that may work at cross purposes. Examination of the interrelationships in the roof of this matrix identifies possible suboptimization of the system as a whole due to the combined effect of "optimized" subelements. The synergistic activities identify opportunities for potential improvements through enhanced interactivity communications.

Either the functional areas or the activities can then be deployed to specific assignments for organizations, groups or individuals. Depending on the size and complexity of the organization, this might be an extension to one of the earlier matrices. Primary, secondary or supportive roles are noted at this step. Overloads and conflicts as well as groups or persons with no responsibilities become apparent. Going to the roof identifies conflicts or contradictory assignments. One benefit of this analysis would be to reveal where additional resources may be needed and to point out where the additional resources may be available. The study can show where non-value-added activities may occur. It can also be used in performance planning and reviews, the development of training requirements and plans and the improvement of communications, especially to access-controlled computer systems.

Further quality system development arising from either the functional area or activities matrix is pursued through another matrix showing the operating practices, policies and procedures that support those activities. By linking through the other matrices, it ensures complete and adequate coverage of the TQM, ISO or other requirements. Omissions and duplications are readily apparent at this point. Often, this can serve as an intermediate step before assigning activities to persons or groups. It identifies which individuals or groups "own" particular operating procedures and policies and have the responsibility for keeping them current and consistent. This is basic to understanding which areas are responsible for what information and provides the basis for development of an effective ISO Document Control Plan.

There are many advantages to this overall QFD-based approach. First and foremost, it provides some structure, discipline and a format for identifying, analyzing and documenting a comprehensive review of the entire quality system. It forces "systems thinking" more than any other methodology we have identified. It provides a means to tackle the task so it can be understood by the entire organization, with roles and responsibilities clearly noted and defined. It also produces, at each stage, a graphical output that assists communication to and understanding by the organization. *The entire effort can and should be tailored to each organization, its internal resources, its size and its scope.* Experience has shown that it is applicable and appropriate for organizations from job shops to large, multisite operations.

The development and completion of the matrices effort normally force beneficial communication between the various affected individuals and groups; in addition, interfaces, interrelationships, dependencies, gaps and inconsistencies are revealed. Maintenance of a consistent ISO framework over time and following a strategy of adding future new requirements or strategies to the most closely linked existing ISO element instead of adding elements or creating a new layer of requirements or activities will immediately show the impact of the new addition to the existing system. This is most easily revealed through use of a roof-type correlation analysis. Resource requirements can be generated from specific activities, tasks and responsibilities expressed in the matrices. Budgets and timelines can become more logical and fact based.

An important benefit of this approach is that it documents the overall quality system and quality plan. It provides fundamental documentation of management responsibility and system definition required by the ISO 9000 series and other industry or customer requirements. It also links the organization's quality manual, policies and procedures, operating practices and work instructions back to the TQM, ISO, customer and other requirements. Policies, procedures, operating practices and work instructions that cannot be linked to a requirement or strategy should be questioned and probably eliminated. This information could be shared with customers or third parties, such as quality systems registrars, and makes the wasteful recasting of spurious "quality manuals" unnecessary. Resources normally targeted in this direction can be redirected to more value-added areas.

When confronted with the need to make changes or to incorporate some new or revised requirements in their quality systems, organizations often fail to take a holistic system perspective and do the necessary upfront analysis and definition. This results in unproductive, confusing and often wasteful activities. Examples usually can be seen where companies independently address TQM, ISO or customer requirements in a disassociated or disjointed manner. The tools and techniques associated with QFD provide a comprehensive, effective approach to strategically and simultaneously address TQM, ISO and other requirements and to show with clarity the relationships between the elements that allow those requirements to be met. Such an approach also avoids turmoil and waste of time, people and resources. The singular application of these tools can provide significant benefits to most organizations.

15

QFD AND CONTINUOUS IMPROVEMENT

Many organizations today are involved in more than their traditional responsibilities: They are also designing, implementing, monitoring and maintaining continuous improvement initiatives. These initiatives go by many names, such as Total Quality Management (TQM), Continuous Improvement Initiative (CII) or Continuous Process Improvement (CPI). The term CPI will used here. Some of these initiatives have been quite successful and have more than met their organizations' expectations. Others have consumed a lot of organization resources without returning as much as was originally expected in the way of benefits. One key to success is to design the CPI initiative around the specific needs of the organization and implement it in a way that fits how the organization operates. This is not to say that the organization will not have to change some of its culture, but much of the success of new initiative' can be attributed to how well it fits the individual circumstances of an organization.

For example, CPI would look different if it were being designed for a small- to medium-sized insurance company than for a large construction company. Besides being smaller, the insurance company would have an organizational structure with a central office and many persons operating by themselves in the field with little or no oversight. There might be several different insurance products that change slowly over the years, and these products would compete with a wide variety of other insurance products, perhaps even other financial instruments. The one-on-one interaction of the field persons with their customers may be the most critical aspect in the case of the insurance company.

On the other hand, the construction company might maintain a sizable central office, but the real action would be the construction sites, where groups of company employees continually interact with each other and to a lesser extent with the customer or the customer's representatives. In construction, the product is usually unique for each site, accomplished on a project-by-project basis, with many projects competitively bid against other companies and some awarded outside of the competitive bid process.

These two organizations would look at their administrative support activities and operations activities differently and would define core processes differently. They would have different priorities both on what to address first and in their expectations of outcomes. There are many different approaches to CPI. Which would be the correct one for each of these disparate organizations? Probably no single approach is correct and would serve them both well. At least not without a lot of changes and adjustments so that it would fit the organization and provide the benefits expected. So we need to ask: How would each of these companies (or any organization) go about designing, implementing and maintaining a CPI initiative that is appropriate for their situation and gives them the outcomes they feel they need to justify the effort?

A key part of the answer is to define the wants, needs, expectations and requirements of the organizations' customers, both internal and external, and then use the QFD methodology to design a CPI initiative that is appropriate for the situation. In talking about what is "appropriate for their situation" and "the outcomes they need," it is essential that it has been established what business they are in, that is, what products and services the organization provides to the customer and under what circumstances (the organization's mission, values and core competencies) and its goals for the future (vision) (see Figure 15-1, blocks A and B). Once the bases of the organization's operations are determined, then planning is done to carry out these operations (see Figure 15-1, block C).

The typical macro steps for moving an organization to the continuous improvement mode are

(a) Understand the current situation, the *AS IS*.
(b) Determine the desired new state, the *TO BE*.
(c) Design the *TRANSITION* from the *AS IS* to the *TO BE*.

The assessment activities discussed below and information from many sources, including the organization's suppliers, internal customers, (em-

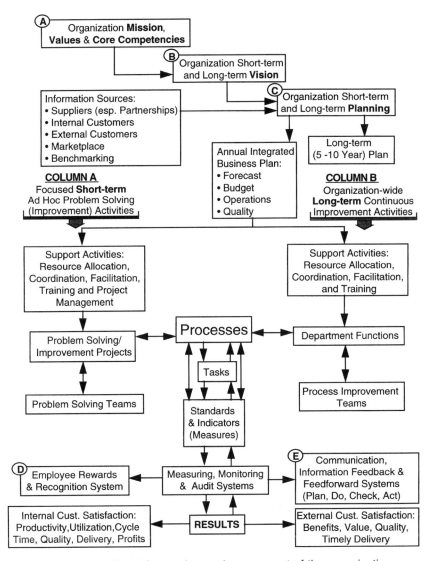

Figure 15-1 Model for continuous improvement of the organization.

ployees) and external customers, help to define (a) above. After gathering inputs from the internal and external customers, QFD can assist in (b) and (c).

The first step in designing a CPI initiative is to assess where the organization is, the AS IS. Areas of assessment would include a high-level view of internal and external personnel regarding where the organization is relative to its competitors. At a lower level, assess what skills the workforce needs and how effective the training system is at providing them. It is essential to understand the strengths of the organization as well as the weaknesses (even to the point of listing and prioritizing the organization's problems, which is very useful if CPI is begun using a focused short-term ad hoc problem solving approach (see Figure 15-1, column A). Areas of consideration include communication, rewards and recognition, team activity, quality of customer interface, cooperation between functional areas (departments), executives' decision -aking modes, performance evaluation system and the current key measures and indicators used to manage the business, to name a few. Assessment instruments based on the Malcolm Baldrige National Quality Award are often used. Some examples are GOAL/QPC's *Malcolm Baldrige National Quality Award Self-Assessment Workbook* and *The Quality Measure* (*Le Qualimètre*) available from le Mouvement Québécois de la Qualité, Montréal, Canada (available in English and French).

Assess the organization on a diagonal "slice" through the levels of the hierarchy, interviewing approximately the same proportion of personnel at each level. In organizations with less than 75 employees, all of the senior managers should be interviewed and as many as a fifth of the remaining staff. Summarize the responses into a report showing where, for example, there is agreement and differences between functions and levels. Conduct a management and peer review of the summarized results, with the different levels and functions reviewing the results separately. Involve both interviewees and noninterviewees in the review groups. Have them answer the question: "Does this summary reflect the organization as you know it? If not, what adjustments would you make so it more accurately reflects the organization?"

Assessing the organization to understand issues and culture before starting to design an appropriate CPI initiative is the same basic approach taken when a product or service is being designed and Voice of the Customer information is collected. Based on the final version of the assessment summary, use QFD to design a CPI initiative that "fits" the organization and draws from the expectations of its stakeholders and customers. QFD can then be used to create the CPI implementation plan and the CPI monitoring and maintenance plan (see Figure 15-2).

Very often management comes to the conclusion that an organization needs to start CPI because of several problems that are making it difficult to maintain or improve its presence in the marketplace. The

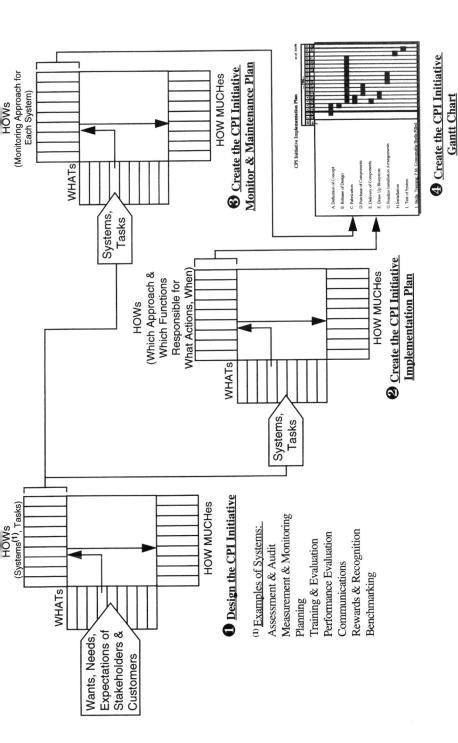

Figure 15-2 Designing the organization-wide long-term CPI initiative.

organization may start CPI in the short run to address these pressing problems (see Figure 15-1, column A). Typically, the organization uses focused, cross-functional teams and problem-solving skills to return areas of subnormal performance to their expected levels. It is essential that at the end of each of these problem-solving efforts a monitoring system be installed as a means of preventing "backsliding" toward the previous situation.

As team-building and problem-solving skills are learned and absorbed by different parts of the organization, the need to design a more broadly based, follow-on approach that can address the issues beyond the initial problem-solving efforts becomes apparent (see Figure 15-1, column B). This long-term approach builds on the knowledge acquired during the earlier problem-solving efforts but expands the number of areas addressed across the organization. Besides addressing specific problems, it is concerned with broader, organizationwide issues such as changing the organization's (a) rewards and recognition system, (b) communications system and (c) planning system (see Figure 15-1, blocks C, D, and E).

Note: Whether the organization starts CPI with specific problems (see Figure 15-1, column A) or with efforts in department functions (see Figure 15-1, column B), the approach deals with the affected processes. That is, CPI is *always* process focused.

QFD should be one element of many in any organization's overall continuous improvement process. One advantage in using QFD to design the CPI initiative is that it requires information from the customers of the initiative. Many of the persons in the organization who are going to be impacted by this new way of doing business are going to be involved up front when they participate in the interviews for the assessment. Their responses will affect how the CPI initiative is structured and implemented. Although in many instances they may not be in favor of changing, at least they are going to be involved in determining many of the elements of the new operating environment. They will also participate in detailing the expectations or outcomes to be derived from the new environment. Due to their involvement in these early activities, there is a greater likelihood that the participants will accept and embrace the CPI initiative. And there will be less need for modification and rework of the CPI system later.

The initial emphasis of a CPI initiative is typically a balance of

1. *Behavioral issues*—establishing a culture that expects and supports change

2. *Analytical issues*—dealing first with the effectiveness issue (doing the right things, those tasks that are viewed as value added by the customer), not the efficiency issue (doing all tasks better or faster, whether they are value added or just cost added)

The planning must be *results oriented* (not activities oriented) and provide for *running the business while improving the business.* In creating the CPI plan, although activities such as the training of personnel and creation of cross-functional teams will be parts of the plan, the basis for success depends on attaining results. To accomplish that end, it is necessary to establish appropriate indicators such as total cycle time and first-pass yield and related goals such as 15 min total cycle time for responding to customer information requests (insurance company) and 98% first-pass yield on architectural drawing approval (construction company) .

CPI must not take on an activities orientation with goals similar to the following because doing more of or concentrating on these activities does not necessarily lead to improvement:

(a) "More training," or "training for skills in the areas of x, y and z"
(b) "More teams," or "teams to deal with problems in areas a, b and c"

An exhaustive plan will address these activities as well as rewards and recognition, communications, planning, and so on (see Figure 15-1, blocks D and E). Some of the elements to consider in a QFD design effort follow.

The rewards and recognition system needs to be modified to recognize, reinforce and balance the new importance of teamwork along with the traditional emphasis on individual effort. Internal customers can give the input necessary to design a system that is fair, easy to administer and rewards the behavior necessary for CPI to be successful. The rewards and recognition system can be redesigned using QFD and should be linked to the level of success each person experiences as determined by the indicators established for each task in the processes studied.

Any existing communications system needs to be augmented (redesigned) so it disperses information throughout the organization about improvement activities, projects, goals, the personnel involved, the methods they used and results obtained. A complete communications system will highlight the key indicators used by different areas to gage their success in the new continuous improvement culture. (This would be the first step in initiating Policy Deployment if the organization is interested in eventually doing that.) The communications system can be augmented using QFD.

The planning system that is used by the organization is especially important because it is necessary to gather information from many different sources and integrate them for a plan that simultaneously addresses (a) business and operating issues, (b) quality issues, (c) expected (forecast) market demand, (d) resources available and (e) their deployment for maximum effect. QFD can be used to design the planning system as well as the actual plans themselves.

QFD can assist in the design of the initial short-term plan as well as the long-term plan. No organization has the resources, money or personnel to do everything at once so it is vital to plan the CPI to the greatest advantage of the organization's situation. In cases where it is possible, the focus should be on solving problems that will yield cost savings in the short term. Once these initial solutions are in place, the funds generated from these cost savings can be used to finance further improvements that take longer to come to fruition.

Another approach is to plan early to attack the critical problems where improvement can be both seen and felt (see Figure 15-1, column A). "Seen" here means that persons not directly associated with the improvement efforts will still be aware of them and their results. If the results have positive effects, there will be increases in employee morale and the process will begin of getting the buy-in from persons in the organization who doubt the effectiveness or necessity of an improvement initiative. "Felt" refers to reduced costs (including reduced costs of quality), increased customer satisfaction and increased revenues and profits.

A high-level model for continuous improvement of the organization is shown in Figure 15-1. Flow charts of the lower level details typical of a CPI initiative (with the various steps organized into three phases to be implemented over $3\frac{1}{2}$–5 years) are shown in Figures 15-3a through 15-5b.

Once the CPI initiative is underway and process improvement teams are formed, QFD is used in the improvement (redesign) of specific processes. To determine the initial group of processes to improve, internal and external customers should be surveyed to determine the importance ranking of the organization's key processes. As a part of the survey, questions about dissatisfaction with specific outcomes from these processes should be included. Then a list of the most important processes in the organization that have the highest degree of dissatisfac-

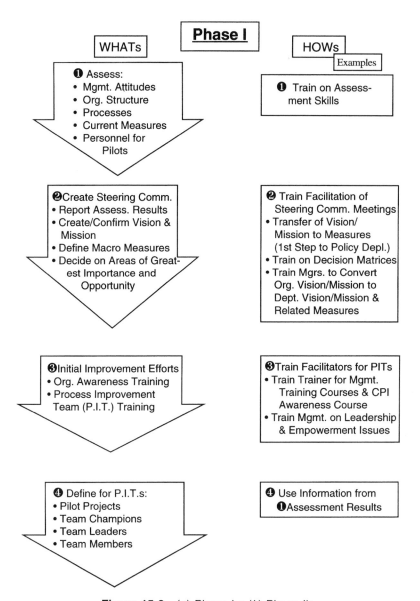

Figure 15-3 (*a*) Phase Ia; (*b*) Phase Ib.

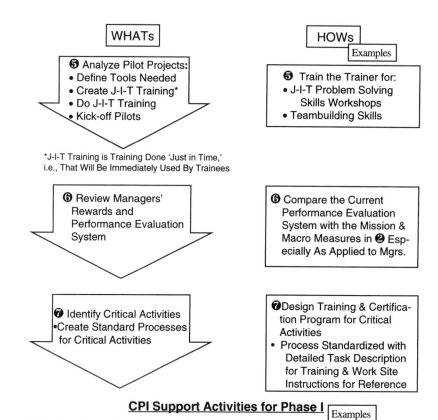

WHATs

❺ Analyze Pilot Projects:
• Define Tools Needed
• Create J-I-T Training*
• Do J-I-T Training
• Kick-off Pilots

*J-I-T Training is Training Done 'Just in Time,'
i.e., That Will Be Immediately Used By Trainees

❻ Review Managers'
Rewards and
Performance Evaluation
System

❼ Identify Critical Activities
•Create Standard Processes
for Critical Activities

HOWs   Examples

❺ Train the Trainer for:
• J-I-T Problem Solving
  Skills Workshops
• Teambuilding Skills

❻ Compare the Current
Performance Evaluation
System with the Mission &
Macro Measures in ❷ Esp-
ecially As Applied to Mgrs.

❼Design Training & Certifica-
tion Program for Critical
Activities
• Process Standardized with
  Detailed Task Description
  for Training & Work Site
  Instructions for Reference

CPI Support Activities for Phase I   Examples

•Design Communications Plan for Whole Organization:
  • Design Communications Plan for Department and Program
  • Create Department and/or Program Newsletter
  • Initiate 'Brown Bag' Meetings with Top Executives, Line Workers and Supervisors
  • Integrate CPI into Current All Organization Newsletter
• Best in Class (BIC) Tours (as Preparation for Benchmarking)
• Awards Ceremonies for:
  • PIT Team Recognition
  • Excellence in Day-to-Day Operations
• Surveys:
  • Of Completed PITs on Needs for Training and Support (to Improve on Next Time)
  • On Organization Training Needs and Feedback on Training Given
  • Addressing External Customer Service Issues (End users)
  • Employees (Internal Customers, Field Service)

Figure 15-3   *(Continued)*

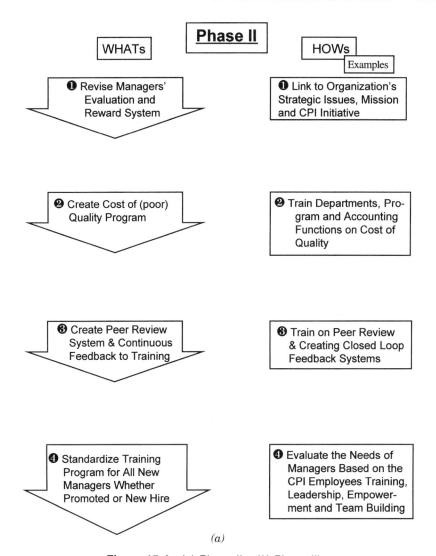

Figure 15-4 (*a*) Phase IIa; (*b*) Phase IIb.

tion is generated. The items listed should be addressed by the initial process improvement teams. As part of the QFD study for improving a process, it is necessary to establish indicators that can be used to monitor and reflect the continuous improvement of each key process. To aid lower level employees in focusing their efforts on value-added tasks, these indicators should be be broken down to lower level tasks, as

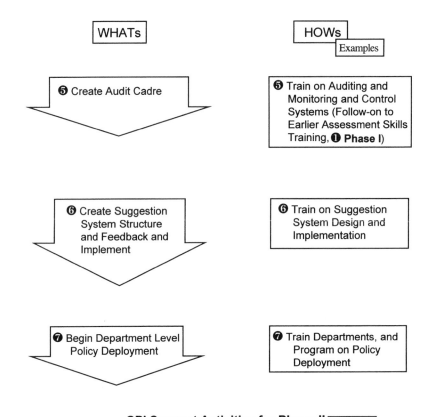

WHATs

HOWs | Examples

❺ Create Audit Cadre

❺ Train on Auditing and Monitoring and Control Systems (Follow-on to Earlier Assessment Skills Training, **❶ Phase I**)

❻ Create Suggestion System Structure and Feedback and Implement

❻ Train on Suggestion System Design and Implementation

❼ Begin Department Level Policy Deployment

❼ Train Departments, and Program on Policy Deployment

CPI Support Activities for Phase II | Examples

• Team Investigates Design & Implementation of a Benchmarking System
• Surveys:
    • Standardize Content, Frequency, Feedback for Customer & Employee Surveys
    • Create Initial Surveys for Suppliers including Partnership Issues

(b)

Figure 15-4 *(Continued)*

is done when applying Policy Deployment. These task-level indicators should serve as inputs to a rewards and recognition system.

Structured correctly, that is, with the larger entities broken down into smaller parts, this process should allow the operations of the organization to be more easily benchmarked (as a result of an enhanced ability to compare the same or very similar tasks between different organizations, i.e., comparing apples to apples). The competitive analy-

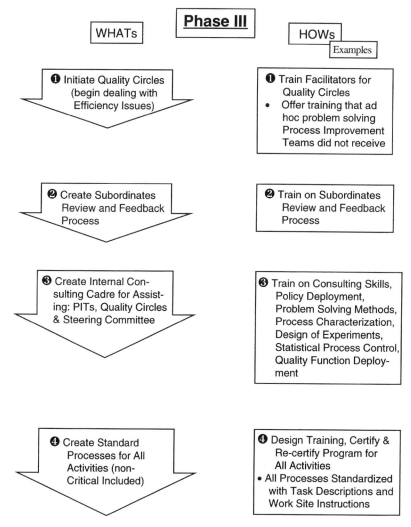

Figure 15-5   (a) Phase IIIa; (b) Phase IIIb.

sis "room" of the House of Quality (HOQ) can be used for benchmarking specific operations with those of other noncompeting organizations.

The importance rankings (for the listed key processes) are provided by senior management and different customer groups. Senior management will also decide which customer groups are most important. Prior to this critical decision-making process, which will give direction to the organization's efforts into the future, it is important to get senior

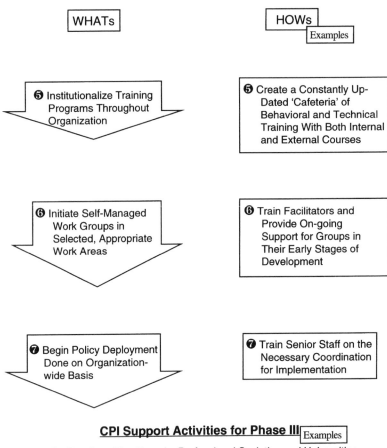

WHATs

HOWs
Examples

❺ Institutionalize Training Programs Throughout Organization

❺ Create a Constantly Up-Dated 'Cafeteria' of Behavioral and Technical Training With Both Internal and External Courses

❻ Initiate Self-Managed Work Groups in Selected, Appropriate Work Areas

❻ Train Facilitators and Provide On-going Support for Groups in Their Early Stages of Development

❼ Begin Policy Deployment Done on Organization-wide Basis

❼ Train Senior Staff on the Necessary Coordination for Implementation

CPI Support Activities for Phase III [Examples]

- Promoting the Benchmarking Done by Professional Societies and Universities
- Benchmarking With Other Non-Competing Organizations
- Awards Ceremony for Quality Circles and Self-Managed Work Groups
- Surveys:
    - Customers (Past, Current and Future)
    - Employees (Past, Current and Future)
    - Suppliers (Past, Current and Future)
    - Other Agencies (Regulatory, Certification, etc.)

Figure 15-5   *(Continued)*

management to agree on a set of criteria (usually linked to the organization's mission and vision) before any discussions about specific industry groups or customers occur. Some of the results coming from this analysis may be surprising.

For example, criteria might address which customer groups are the largest current sources of an organization's revenue and which market segments currently contain concentrated resources. Sometimes it happens that the organization is effectively ignoring its current largest sources of revenue while trying to develop the next two likely (but not yet anywhere near) large sources of income. Some criteria for evaluating customers might be about past revenue or profits and likely projections of future revenue or profits (based on loyalty/partnering, technology, geography, etc.), technology linkages, market developments, distribution linkages, increased exposure in global markets, as well as ease and cost of field support, and so on.

The results of the senior management ranking of the importance of key processes are then compared with the various customer groups' rankings of importance following which a mechanism for resolving the differences is set up. Preferably, this would be a neutral (nonpolitical) way of getting senior management to begin to see how the importance rankings of their most important customer groups (based on the senior management criteria) are linked to their (senior management's) goals, whether they be greater market share/penetration, return on investment, profits or whatever performance measures are desired.

For each identified disparity (gap or delta), a matrix showing how to address the disparity, give measures(indicators) and identify resources related to the HOW is created.

Assumptions should be stated at the beginning of the CPI initiative (mission/vision, short-, medium- and long-term goals and/or SWOC-Strengths, Weaknesses, Opportunities and Constraints). A Pareto chart should be created of the customer groups' current importance and future importance to the organization while reaching its mission/vision and stated goals. Then, using matrices, it is important to show how each customer group will be addressed in the CPI initiative to propel the organization to its goals.

After the initial assessment of a wide range of customers and with senior management looking at where the greatest benefit could be derived from the allocated resources, the insurance company and the construction company each chose a different focus. The insurance company focused on (1) customer surveys and feedback and (2) Exciting Quality, that is, establishing new, innovative products as well as new ways of marketing, as the top two most important issues for its CPI initiative. The construction company targeted (1) job costing and monitoring and (2) supplier partnerships as its top two initial efforts.

In addition to aiding the CPI process for an entire organization, QFD can assist with doing continuous improvement at the product or service design level. It can be used to keep a running record of the competitors' latest offerings in the market. The competitive comparisons section of the A1 matrix can be changed as new offerings are introduced, and this provides graphic evidence of the gaps in features and functions between the organization's current offerings and those of the competitors as they develop (see Figure 15-6).

The sales or marketing function can accumulate the latest information from the marketplace on the new features and functions being offered and keep an updated, uncontrolled version of the A1 matrix. (Document Control should always keep a controlled copy of all the QFD matrices that apply for each product or service as it was released or introduced.) The frequency of review of these gaps would depend on how fast the industry is changing. Providers of products and services such as pizzas (and the related home delivery service) might review the gaps every six months, domestic refrigerator ice makers every year, whereas designers and makers of notebook computers might review the gaps every month.

In fast-changing industries, regular updates of competitive comparison graphs should be followed by immediate joint meetings with engineering and manufacturing. The effects of the competitors' new offerings, how they link to the customer's desires for certain benefits and what opportunities there might be to augment the organization's current products or services to address the same benefits are discussed. (That is, a feature/function that is an extension of the current offering would be quickly offered as opposed to a complete redesign.) It is important to be on the lookout for situations in which new competitor offerings have features or functions that do not link to known customer requests from previous market information gathering. When there is no linkage between a feature or function and known customer demands or requirements (the WHATs), this may be an indication of a new trend. If so, marketing/sales will need to investigate and inform senior management.

Continuous updating of the A1 matrix competitive comparison graph (see Figure 15-6) gives advance warning of large gaps that need to be addressed in the next generation of a product or service. Obviously, taking the competitive offerings already in the marketplace and recording them on the competitive comparison chart leaves the organization in a reactive mode, but at least new benchmarks are identified as they occur. In addition, it gives a convenient starting point for any discussion on augmenting the current product/service, either to "tweak" it and introduce an interim response or to provide the basis for the next generation of product/service.

If briefings by marketing/sales are held on a regular basis, with apparent trends identified, engineering design and manufacturing will gain a better understanding of how much they must "lead" the market

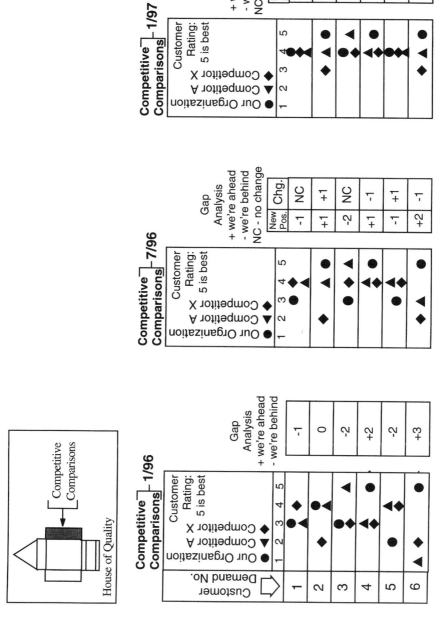

Figure 15-6 Monitoring the market with the A1 matrix competitive comparisons.

to gain competitive advantage, in terms of both cycle time to volume production and levels of performance enhancement. A clear understanding of the trend will often allow a design team the ability to design (or build in) future modular add-on capability. Certain telecommunications systems, computer networking systems and automobile engine control systems are examples of understanding trends and designing in future capabilities.

16

QFD AND OPERATING ENVIRONMENT AND PRODUCT LIFE CYCLE

Whether QFD is used for a completely new design or to update a design for the next generation of a product, there are external factors that must be integrated into all design activities on an ongoing basis. A company with a structured product planning process needs to create and maintain a foundation of knowledge about the environment in which the company and its products operate. This foundation of knowledge needs to address all of the aspects of the product from "womb to tomb." It is common to focus design activities on the beginning of the product's life, making it easy for the customer to install, set up and integrate the product into his or her existing operations/life-style. However, design must also address issues such as safe disposal/recycling at the end of the product's life. We all want the advantages of smoke detectors and automotive air bags, but at the end of their life, those products represent a toxic risk and should be disposed of properly. We need to design those products in a manner that assists the disposal process.

How can a company mesh the changes in the operating environment with the product life cycle? (Note that in some instances the changes are so sweeping that they apply not just to a particular product but also to a product line or to a company's entire operations.) QFD can be used to raise and maintain an awareness of all the operating environment issues, to devise technological responses to them and to assist in integrating these responses into product designs as they occur.

There are several areas where the outside environment impinges on product design activities. Government regulations are only one area of concern. As the areas of society, community and industry that are regulated increase and these regulations cover a greater expanse of human activity, they can be viewed either as an impediment to company goals or as an opportunity. If methods like QFD are used proactively, these lemons can be turned into lemonade. An example is the early regulation on smokestack emissions in the US, and the resulting products necessary to scrub the chemical and particulate emissions. Today the United States is the world leader in this technology, which has evolved into a major export sector for the United States. However, without the initial "push" from government regulations, it would not have happened.

There are other major pushes. The American Disabilities Act (ADA) has fostered a variety of products to meet the needs of disabled persons. Germany has regulations requiring manufacturers to address the issues of product disassembly as well as component and material recycling. Firms that intend to be global marketeers will need to stay ahead of trends in regulations in their target markets. Other examples are the push to reduce air pollution caused by lawn-and-garden equipment in the United States and noise pollution at airports everywhere. European governments have passed a series of regulations affecting noise levels at night for airports. These regulations have directly impacted the design of the next generation of Auxiliary Power Units (APUs) used in aircraft. The APU is the source of electrical power when the main engines are off and the aircraft is being serviced. Since most aircraft receive their routine maintenance and cleaning during the night, it is essential that they are equipped with an APU that is quiet enough to be used.

In the design of each cycle (generation) of product the design team should be aware of regulatory trends, anticipate new regulatory requirements and create technical responses to them. These are important not just to satisfy the regulations from a legal point of view once the regulation is actually in effect but also to judge the effectiveness of the technical answers to the defined problem before the law is passed. In some instances the regulatory-prescribed answer to a problem is poorly conceived and requires technical review. Such was the case when sealed beam headlights were regulated into existence. The Europeans, who did not have such a regulation, had much more powerful and more precisely aimed headlights that gave superior road illumination decades before Americans were legally allowed to use similar designs.

In some cases, technically valid approaches do not adequately address the problem when the product is put into volume production. This occurs because the production cannot be well controlled and is not consistent as well as because of unexpected or unintended actions by customers (or others) in the real world. For example, ABS brakes have not met their expected goals of accident reduction because drivers "know" they have to "pump" their brakes when they are skidding. But ABS brakes are designed so the driver pushes and holds down on the brake pedal and the ABS computer does the rest, pushing and releasing the brakes hundreds of times more often than a human could. With some focused education of the driving population on how to use ABS brakes, they may yet reach their potential for accident reduction, but this is one element of the design environment that was not anticipated.

Understanding the operating environment can be as insidious as just acknowledging the demographics of the market. The increasing number of senior citizens make a good market. Studies show they have much greater assets and buying capacity than previous generations. But efforts to design for this market must recognize critical aspects of these customers up front, even before the design process for a particular product begins. This large and growing segment does not, as a rule, see or hear as well as the general population. So they require more light to have the same recognition of conditions (the can opener knife is not fully retracted upward, so the can cannot be removed from the can opener, but the knife is under a shaded overhang with no other source of light than the overhead kitchen lights so the person cannot see the condition and push the actuation arm upward).

Senior citizens are experienced and knowledgeable but have less strength (and often less feeling) in their hands, wrists and fingers. As a consequence, "kid-proof" medicine bottles are also senior citizen proof! The setting, maintaining and use of electronic devices needs to be more user friendly and less frustrating. Some devices have sequences that are neither simple nor intuitive. Creating a methodology for designing these sequences would build a foundation of knowledge useful to many products, even families of products. This design would need to address how humans "learn" to set up and operate a device. This work would not necessarily be unique to a particular product. It would raise the awareness of designers to the demographic environment, which is vital for success if that is one of the target markets.

Besides the somewhat lessening of senses and strength, the senior citizen often uses a product or service less often. During times of storage/nonuse there should be no deterioration in the product's functional capability, and when the product is used again, it should be easy for the user to remember how to use it. Consumables in the product should be easily accessed and changed out, and the whole process should be intuitive (e.g., the labels on the consumable should be right side up, in the correctly installed orientation). Small, difficult-to-see and difficult-to-feel levers, grooves, and so on, should be avoided.

For any product that the company is thinking of designing or redesigning, a foundation of information relating to the environment in which it will operate should be established at least three months before beginning the product design exercise. Current regulations must be known, and if at all possible, the trend in regulations should also be known and detailed. In many fields there are traditional regulatory leaders. For example, in the automotive and general pollution fields, California seems be the trendsetter. If there is not time or resources to talk to a large number of entities about upcoming regulations, then talk to the trendsetters. In Europe, the safety trendsetters are in Sweden and Germany and "Green" politics is German, although the Scandinavians have a tradition of being environmentally aware.

It is essential in products that have long life-cycles that the trends be well understood, not just so a product can have a new feature caused by a regulation that will not be in place for five years but also because it is disappointing to have a product or product line discontinued for lack of some minor provision that, if part of the initial design, would have made meeting the regulation a trivial task. Step changes in the regulatory environment can often be discovered by monitoring the legislative process—but there are no guarantees!

In the case of the gradual changes in the general product operating environment due to demography, there can be no excuse for ignoring and/or not exploiting trends that have been in the making for decades. In other cases, the time leading up to the change was considerably shorter and required greater, focused effort in the affected designs, for example, the change in the refrigeration industry in switching to an environmentally friendly refrigerant and, of course, the change from leaded to unleaded gasoline. Other operating environment trends exist that will affect the design activities of several industries.

Everyone will be affected by the upcoming changes to electric utilities. As the electric generation and distribution utilities are deregulated and electric utilities choose to be more conservation oriented and conservative about building additional electric-generating plants, line delivery voltages will likely vary more and brown-outs will be more frequent. All types of electrical gear will have to address the consequences of these operating environment aspects in their new designs. In telecommunications, as greater portions of the electromagnetic spectrum become allocated, there will be greater interference from spectrum "neighbors." For processes that use solvents that contribute to

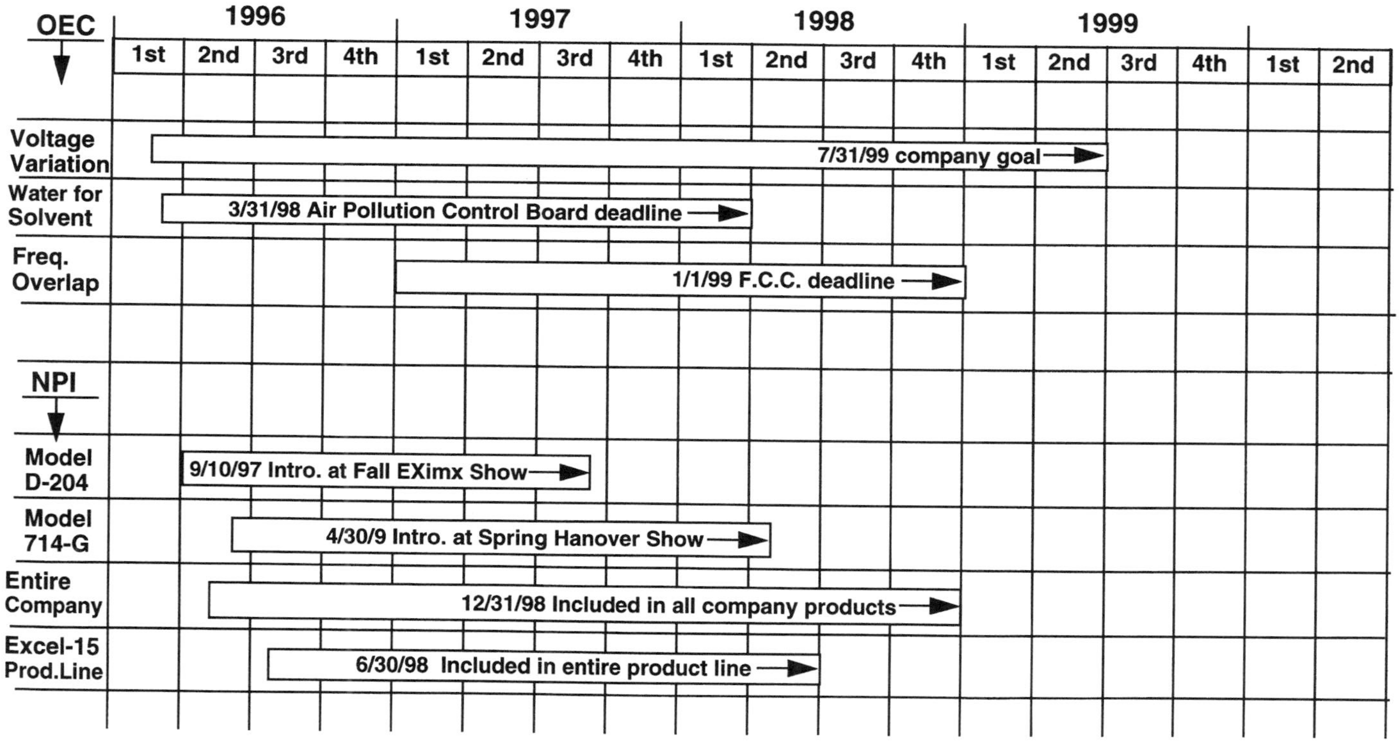

Figure 16-1 Parallel timelines for Operating Environment Changes (OEC) and New Product Introductions (NPI).

air pollution, there is a push to change all processes over to water-based solvents. In many instances, there is an easily substitutable product. For others, it may take years to develop a water-based equivalent.

One way to improve all of your company's design activities is to create a response for each identified change in the operating environment. Detail the current and likely future operating environment changes that will affect your products and their associated timelines, noting whether there is a specific deadline (like the switch to the

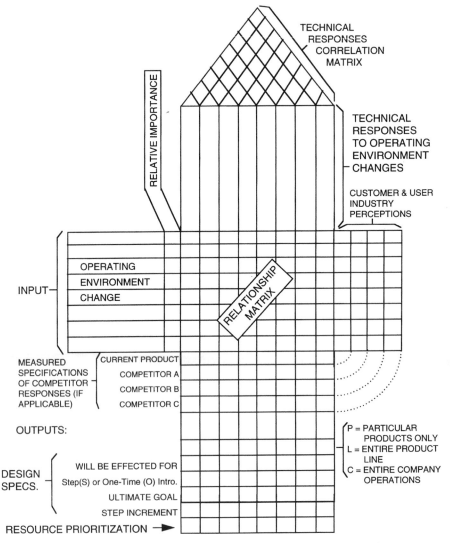

Figure 16-2   Generating responses to the operating environment changes.

environmentally friendly refrigerant) or a gradual switchover (like the switch to unleaded gas). Now, create a timeline showing your current and upcoming product life cycles. Note the likely new product design projects and redesigns of current products. Draw these two timelines in parallel (see Figure 16-1). For those situations where there is a specific deadline, determine if the issue will be dealt with on a product-by-product basis or whether a companywide or product line approach will be taken. Agree on the lead time necessary to create a response to the upcoming change and then create a design team to address the issue.

List all of the changing operating environment factors and use them as inputs (the WHATs) to the QFD A1 matrix (see Figure 16-2). Importance factors in this case will be determined by the operating environment, with safety items most important (likelihood of death or permanent injury), company reputation next and fines and operation suspensions by regulators last. Across the top of the A1 matrix will be the technical responses to the operating environment changes, the HOWs. Once the matrix is complete, look in the roof of the house for any conflicts and resolve those issues first. Coming out of this matrix will be a list of design activities that must be accomplished on a product on a product-line or on a companywide basis and, when accomplished, will serve as inputs to the next product design project where the changing operating environment is applicable.

The operating environment design team should be an on-going team that meets every three months (more often when operating environment issues are changing rapidly) and have an agenda to

- Review the current set of responses to the list of operating environment issues, determine if the responses are adequate and determine if they need to be updated.
- Discuss any operating environment changes that are known to be coming.
- Discuss operating environment changes that are known to be coming but are not on the current operating environment list.
- Appoint team members to work with any ad hoc product design team that needs help integrating the operating environment responses into their product design.

17

QFD AND DESIGNING MEASURING AND MONITORING SYSTEMS

There is a growing realization within modern-thinking organizations of the importance of identifying, understanding, characterizing, measuring and monitoring all of the organization's key processes on an ongoing basis. In fact, with the rapid changes occurring in the environment surrounding the organization, most of which the organization has very little control over, managing operations within the firm takes on added importance. The ability and willingness of an organization to react rapidly and appropriately to situations that could have a negative impact as well as to position itself to seize opportunities as they present themselves are essential for a firm's well-being. The deployment of information systems able to collect data, perform analysis, generate information and then distribute it wherever and in the formats needed to assist decision making is important to an organization's competitive advantage, even for the firm's survival.

Information Technology (IT) has developed to the point that it is inexpensive and implementing good information systems is relatively easy. So the question is not whether to implement a system (i.e., will the organization be able to derive benefits that exceed the costs of the information system?) but rather what information system should be implemented? This design activity can be aided by the use of QFD. In fact, in the manufacturing sector, this issue was one of the two reasons for originating QFD. In the "History of Quality Function Deployment in Japan," Akao (1990a) mentions, "there were two motives that led to the birth of the QFD concept." One was to have a methodology that would assure the quality of the product's design and the other was to be able to create a Quality Control (QC) process chart for the manufacturing floor prior to the initial production run. (Japanese companies had been using QC process charts for some time by the 1960s, but the charts were generated *after* the product went into full production, not before.)

The QC process chart that Akao mentions details the key parameters at each step in the manufacturing process. However, the use of a QC process chart is not limited to manufacturing endeavors. It can easily be applied to the monitoring and control of any process, whether (a) totally administrative, such as the interviewing and hiring of new employees or the processing of insurance claims (e.g., The Budd Company has used QFD for improving its engineering proposal system), (b) an administrative process that is related to production, such as creating the production schedule or Bills of Material, or (c) for actual production operations. Typically QC process charts are organized by process steps arranged in chronological order. For each step, the tasks to be completed, the characteristics involved and their related parameters are identified. For each parameter, the QC process chart details who will monitor that parameter, the monitoring method to be used (the instrument or measuring system and procedure are noted), the acceptable range of responses and the reaction plan should a response fall outside the acceptable range. For an example of such a QC process chart (also called a Control Plan) see Figure 17-1.

In the American Supplier Institute (ASI) Four Phase approach to QFD (see Figure 17-2), the first matrix (the Product Planning Phase) translates customer requirements into technical design requirements for a product. The second matrix translates technical design requirements into component part characteristics. The third matrix translates component part characteristics into manufacturing production operations needed to meet those characteristics. The fourth matrix translates manufacturing production operations into production requirements. The production requirements are stated in terms of the parameters that need to be controlled to assure that production operations will produce a product/service that meets customer requirements. To assist in designing and deploying a measuring system that addresses all of these parameters, a fifth matrix could be added to the first four (see Figure 17-2). This Measuring Planning Matrix would be used to define the measuring system needed to assure that the parameters listed in the fourth matrix are effectively monitored, that the data collected accurately reflect the production process and that the data are quickly and easily available to assist timely decision making.

The WHATs or inputs to the Measuring Planning Matrix would be the production requirements parameters. The HOWs would specify a

Control Plan

Control Plan Number 002 | Prototype ☐ | Pre-launch ☐ | Production ☒ | Key Contacts/Phone T. Smith/313-555-5555 | Date (Orig.) 3-1-97 | Date (Rev.) 4-20-97

Part Number/Latest Change Level Circuit 10/8 — Core Team — Customer Engineering Approval/Date (If Req'd)

Part Name/Description Electronic Circuit Board — See attached — Customer Quakity Approval/Date (If Req'd)

Supplier/Plant ACR Control — Supplier/Code 439412 — Supplier/Plant Approval/Date — Other Approval/Date (If Req'd)

Other Approval/Date (If Rec'd)

Part/ Process Number	Process Name/ Operation Description	Machine, Device Jig, Tools For Mfg.	Characteristics			Special Char. Class.	Methods						Reaction Plan
			No.	Product	Process		Product/Process Specification/ Tolerance	Evaluation Measurement Technique	Sample		Control Method		
									Size	Freq.			
2	Soldering connectoions	Wave solder machine	1	Wave solder height			2.0 ± 0.25 mc	Sensor continuity Check	100%	Continuous	Automated inspection (error proofing)		Adjust and retest
			2		Flux concentration		Standard #302B	Test sampling lab environment	1 pc	4 hours	x-MR chart		Segreagate and retest

Figure 17-1 Example Control Plan (also called QC Progress Chart). The "characteristics" column has only one entry per line under either "Product" (R-criteria) or "Process" (P-criteria), not both. The "Special Characteristics Classification" column contains symbols highlighting the critical or safety-related nature of a process step. The symbols used are unique for each industry or organization.

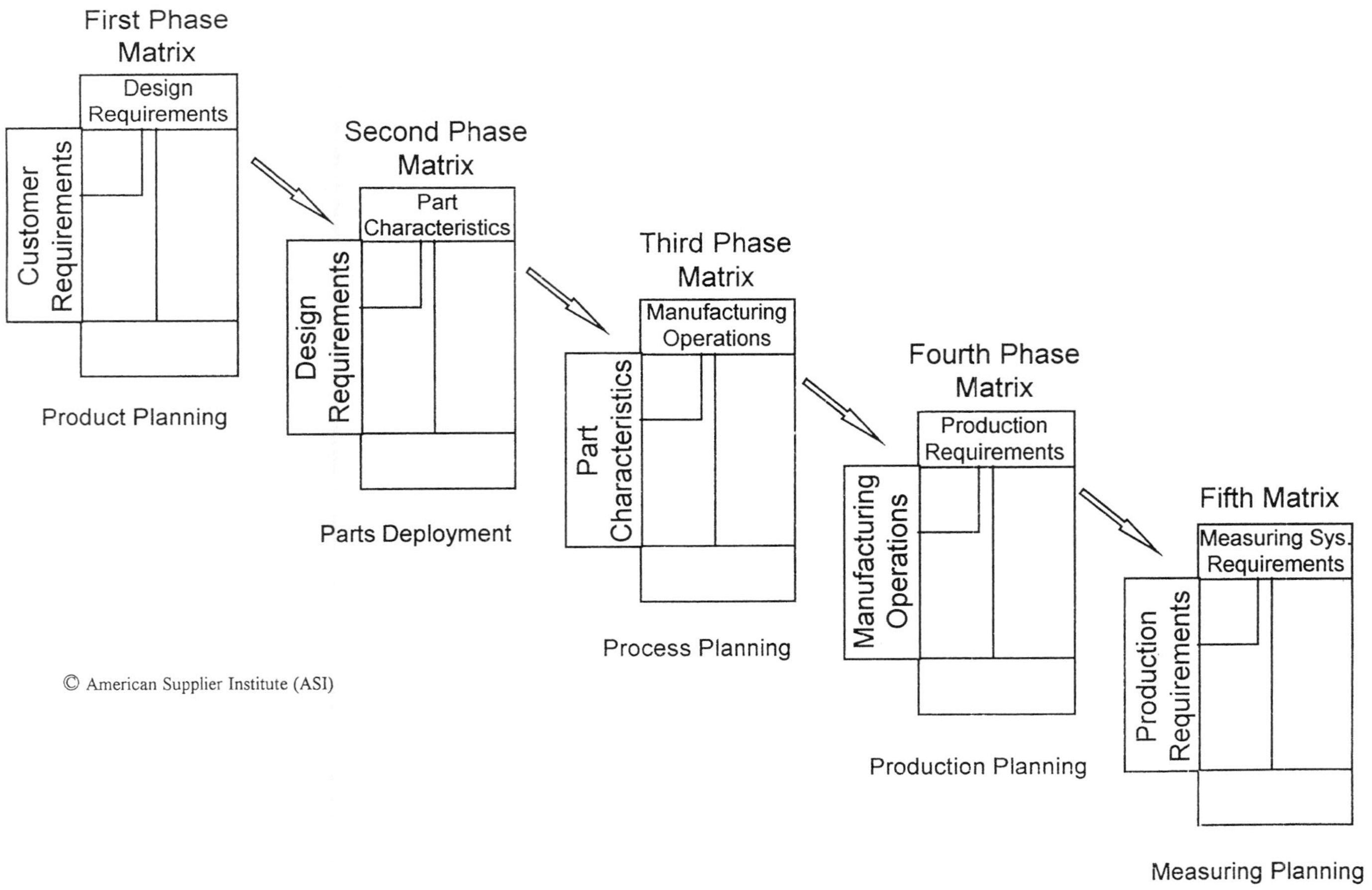

Figure 17-2 ASI Four Phase approach to QFD with fifth matrix added for measuring system. (Copyright by the American Supplier Institute.)

© American Supplier Institute (ASI)

First Phase Matrix — Customer Requirements / Design Requirements — Product Planning

Second Phase Matrix — Design Requirements / Part Characteristics — Parts Deployment

Third Phase Matrix — Part Characteristics / Manufacturing Operations — Process Planning

Fourth Phase Matrix — Manufacturing Operations / Production Requirements — Production Planning

Fifth Matrix — Production Requirements / Measuring Sys. Requirements — Measuring Planning means of measuring and monitoring for each parameter. The production requirements are the parameters that need to be controlled in the production process if the customer requirements listed as inputs (the WHATs) to the First Phase Matrix are to be delivered. To capture and understand all of the elements that affect the design of a production process measuring system, we need to start at the very outset of a product/service design project.

MEASUREMENT SYSTEM DEVELOPMENT (MICROLEVEL)

The typical new product or service goes through five stages from concept to fully ramped production:

(a) Concept Selection
(b) Product/Service Design
(c) Product/Service Confirmation/Test
(d) Production Planning
(e) Production

Using the QFD methodology, the initial one-time activities [(a)–(d)] build a solid foundation for and lead to the major on-going activity (production) accompanied by on-going communication, process measuring and monitoring and organizational measuring and monitoring (see Figure 17-3). Each matrix has elements such that the initial customer requirements can be traced through a series of translations all the way to the actual production process.

To better understand the series of translations, consider an example of a group of customers surveyed on their personal requirements for a car. They might answer they want a car that is comfortable and inexpensive to own. When questioned further about "inexpensive to own," the customers indicate that low operating costs or good fuel economy will satisfy that requirement. In working on the First Phase Matrix and translating the "good fuel economy" requirement of the customer to the technical design requirements, the design team might use several approaches simultaneously: a smaller displacement engine in conjunction with a computer-controlled transmission propelling an aerodynamically styled, lightweight body.

For the second phase, the overall technical design requirements are converted to component parts. In this example, the lightweight requirement could be reached by using a combination of stamped

Measurement System Development Overview

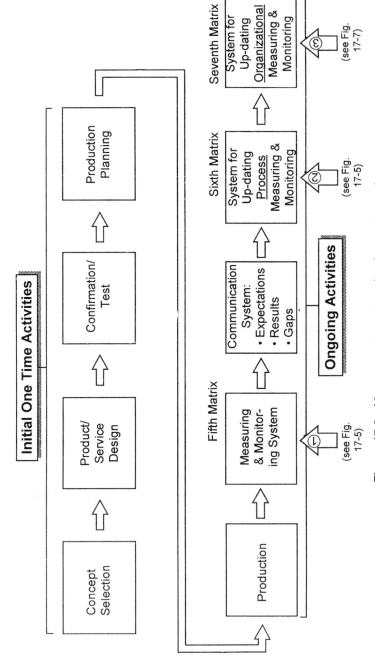

Figure 17-3 Measurement system: development overview.

aluminum and fiberglass body parts. In the third phase, the component parts, say fiberglass fenders, hood and door panels, are translated into a production process during which the fiberglass shapes are created by blowing a mixture of fiberglass particles and resin into a mold. In the fourth phase, the production process is enhanced to include the controls necessary to assure the expected output from the process.

In the case of fiberglass resin, the ratio of the two constituents, the thoroughness of their mixing, time from initial mixing to combining with fiberglass and the temperature at the time of mixing are some of the parameters that need to be controlled to assure the characteristics of high-quality fiberglass panels. These controlled parameters, which are detailed in the Fourth Phase Matrix, can be used to fill in the typical process control plan such as is specified by QS-9000 (see Figure 17-1).

To be considered successful, a process must consistently produce product or service output that meets or exceeds customer expectations. To manage and assure process performance, it is essential to establish monitoring methods and measures that both guide and confirm the production process. It is sometimes useful to take the various elements that relate to measurement and monitoring, which are embedded in a series of QFD matrices, identify and classify them and then create a chart showing the linkages. This aids the understanding of the many different linkages between the customer requirements and the production process characteristics and related measuring system. Note the three step sequence:

1. Establish the characteristics desired from the production process (these can be located either at the end of the process or after a step in the process is completed or both).
2. Detail the parameters that control these characteristics.
3. Create a measuring system that measures and controls the parameters in-process and/or measures the resulting characteristics after the process is complete.

One way to classify these embedded elements is as follows:

(a) C-criteria are high-level needs, benefits or expectations the customer has for the newly designed product/service/process. These are the customer requirements (the WHATs), and our success is gauged by the extent to which we meet or exceed the C-criteria. C-criteria are often stated in customer language or "jargon." Words such as *user friendly, easy to use, flexible, comfort-*

able, long-lasting, and *dependable* are often used by customers to express their desires. These terms, by themselves, do not provide designers or engineers a clear enough picture of the necessary process outcomes or results (which allow customer expectations to be met). They also do not address the elements within the process that if monitored will cause the results to consistently meet expectations. So additional criteria are needed to address end-of-process results (R-criteria) and in-process monitoring results (P-criteria).

(b) R-criteria are the outcomes or results several of which, taken together, satisfy the C-criteria. R-criteria apply to the results obtained after one or several tasks are performed or a process is completed. They evaluate the previous activities and confirm that the desired results have been obtained. The information from checking the R-criteria can be fed back (upstream in the process) or fed forward (downstream) to adjust the process, but only after the fact. Although R-criteria are fairly specific and are stated in technical jargon, there is a better way to control a process than measuring after the fact and then making changes, that is, go upsteam in the process and then identify and use P-criteria.

(c) P-criteria allow monitoring and measuring of operations as they happen and allow for real-time adjustments. Meeting a particular R-criterion usually depends on one or more P-criteria being monitored and in control.

It is possible to define a hierarchical system of P-, R- and C-criteria such that the P-criteria are monitored and assure that the R-criteria are met. Groups of R-criteria, in turn, are monitored, and these results assure that the C-criteria are met. Note that in some cases a single P-criterion might be linked to more than one R-criteria, and likewise for the R-criteria linkages to C-criteria. A simple CRP-criteria relational pyramid is shown in Figure 17-4.

In a typical process, there might be as many as 12 (or more) R-criteria for each C-criterion and as many as 12 (or more) P-criteria for each R-criterion. As data are collected and greater experience is obtained with the process, often the number of R- and P-criteria can be reduced with no difficulty in meeting the C-criteria. This is due to several effects. Some characteristics show very little change over time, so they need to be checked periodically but not monitored continuously. Sometimes two or more characteristics move in tandem, so it is only

Generic CRP Pyramid

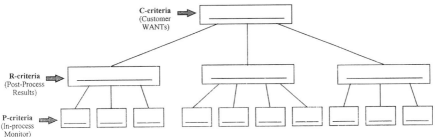

Figure 17-4   CRP pyramid.

necessary to monitor one characteristic. These effects can be detected by observing the data from the monitoring and measuring activities after the measurement system is defined and operational. (It is usually not possible to know the interrelationship of these effects during the design of the measuring system, so it is essential that this analysis be accomplished after the new measuring system has been inaugurated to catch these opportunities for reduction in measuring activities.) However, the same effects could have been detailed through designed experiments (DOE).

In the Product/Service Confirmation/Test stage of new product development, the relationships between C-, R- and P-criteria can be established for those linkages that are historically known. For other cases, DOE should be performed so that the key relationships between various characteristics and the customers' requirements can be established. The DOE can also detail the strength of the linkage between the characteristics and requirements and the acceptable range of parameter values necessary to achieve optimal results.

When creating a measuring and monitoring system, for those requirements customers rate as most important, it is important to concentrate on ways to use P-criteria. They are better able to avoid out-of-specification situations.

In the fifth matrix, which should be used for designing a monitoring and measurement system, three different approaches must be addressed simultaneously. The overall approach should emphasize prevention: preventing the process from producing products or services that do not meet expectations. There are usually known causes for the process producing substandard output; the measurement and monitoring system needs to detect and communicate that information quickly and accurately when it happens. However, because there is always a chance for a lapse to occur in some unexpected way, the system needs to have an information base that can support Root-Cause Analysis (RCA). For example, a system design that is entirely focused on prevention cannot assist RCA efforts when things go wrong. Thus, a leveraged approach that takes the data generated during the regular prevention monitoring efforts and archives it with traceability "tags" is employed to assist in later use of the data should RCA studies be needed.

In creating the fifth matrix, the inputs (the WHATs) are the production requirements. For each production requirement the design team should consider monitoring system elements (the HOWs) that address (a) the prevention, (b) the detection and (c) the investigation issues outlined above. Not all production requirements may be critical enough to have all three addressed, but for the most important all three need to be in place. The importance can be determined from considering (1) the risk of producing substandard products or services and (2) the impact on customers' wants. The allocation of an organization's resources for measuring and monitoring is then based on the customers' importance ratings. The following elements are key to determining the risk and understanding the impact:

(a) *Criticality.* How critical is this characteristic relative to meeting the customers' most important wants?

(b) *Frequency of Occurrence.* How often does this characteristic fail to meet expectations?

(c) *Process Volatility.* How much can this characteristic be expected to change and how quickly?

(d) *Ease and Accuracy of Timely Detection.* Can the change in the characteristic be easily and quickly detected and is the parameter monitoring method accurate?

Following the completion of the fifth matrix and the initial Measuring Planning Matrix, all the linkages between customer requirements (C-criteria) and the measuring system elements (R- and P-criteria) will have been detailed. Then, follow-on planning should be completed to anticipate both changes that can occur as the production process matures as well as the necessary responses to those changes. Any production process is subject to change and will evolve as more knowledge is gained, failures are encountered, "fixes" are put into place and improvements are made. In addition, the marketplace can also be expected to change, thus placing additional needs for change on the product/service and its capabilities and delivery system. These two reasons—change in the process and change in the market—drive a need for a follow-on matrix to assist the ongoing production process in creating an ongoing parallel process for evaluating and updating measuring planning. The sixth matrix assists in the design of a measuring planning updating system that will constantly gage expectations, look at the production process results (quality and character of output) and address the areas where there are gaps.

This sixth matrix is created with the goal of maintaining a monitoring and measuring structure that updates the process measuring system so it is always relevant and effective. This is necessary even though there are changes (improvements) in the process and changes in the external environment (e.g., due to technology and regulatory issues). It does this through constant monitoring of the gap between expectations of the marketplace and the results of the production process. The measuring system focuses the use of the organization's resources based on the customers' importance priorities. Specifically, it should reduce the resources spent on process characteristics that show reduced variability, as demonstrated by higher Cpk indices or lower defects per million opportunities (dpmo), for example, and shift resources to other characteristics with greater variability. Also, the system should shift resources to characteristics that have gained in importance in the marketplace, for example, more emphasis on reduced vibration or emissions.

The measurement monitoring system's design should be dynamic and allow for constant updating as conditions change. Processes can both improve and deteriorate. Process status indices such as Cpk and Cp (process capability index) will reflect these new conditions and should drive whatever approach is taken to monitor the process. Two common approaches used to monitor the process are Statistical Process Control (SPC) and First Piece sampling. (*Note:* First Piece sampling should not be confused with First Article. First Article is part of the qualification of a product and the process that produced it. It is done only once in a particular product configuration's life. First Piece sampling is usually performed at the beginning of each shift or day or week to verify that the output is still within specification.)

Many organizations do not have a specific plan for dealing with reduced risk as variability is reduced. But one of the benefits of reduced variability of one characteristic is the ability to move resources and concentrate them on other characteristics, thereby constantly improving the quality of the overall product/service. Some organizations have established rules of thumb to define when to switch between different modes of monitoring, such as the following. When a process characteristic's Cpk is greater than 1.00 but less than 1.67, SPC would be used. However, the frequency of the sampling performed is greater for a Cpk of 1.10 than for a Cpk of 1.60. For a Cpk greater than 1.67, a First Piece sample approach is usually more appropriate with the time period between samples being longer for larger Cpk values. The measuring and monitoring system needs to establish "trigger" points or thresholds for (1) changing the frequency of SPC sampling, (2) changing the length of the time period between First Piece samples and (3) changing between SPC and First Piece samples.

The measuring and monitoring activities at the microlevel are important because they include the production tasks, activities and processes that give the product or service value to the customer. The overall measurement system development process is shown graphically in Figure 17-5. Microlevel actions are responsible for a large share of the benefits that the customer receives. Because of the linkage to creation of value for the customer, it is essential that the microactions be understood, measured and monitored on an on-going basis. However, the microlevel measuring takes place within the context of the operations of an entire organization. To be effective, all measuring and monitoring systems that are set up to work in a particular area need to interface with a larger system (see Figure 17-6).

MEASUREMENT SYSTEM DEVELOPMENT (MACROLEVEL)

No matter how well designed and implemented, a microlevel measurement system cannot do its job well unless it operates within a well-conceived macrolevel measurement system, the result of an organization that is driven to measure the right things and then manage by these measures. Done well, it is the macrosystem that sets the stage, creates the foundation and framework and reinforces the culture of measuring for the entire organization. The macrosystem involves all the activities of the organization and monitors how well the organization is meeting both day-to-day targets and longer term goals. The microsystem works on the individual processes that support these targets and goals. These processes may be involved in the actual production of a product or service, general administration processes (such as accounting and finance) or administrative processes in direct support of production (such as purchasing and human resources).

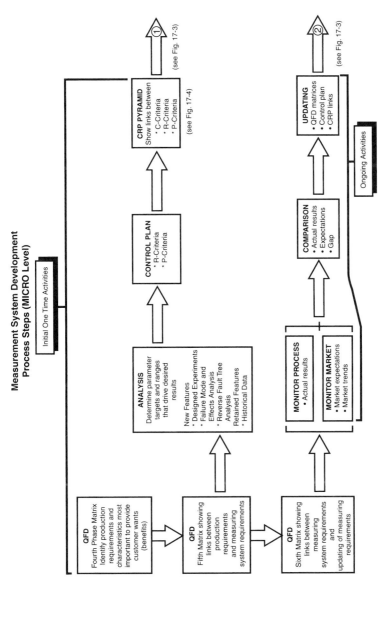

Figure 17-5 Measurement system development process steps: microlevel.

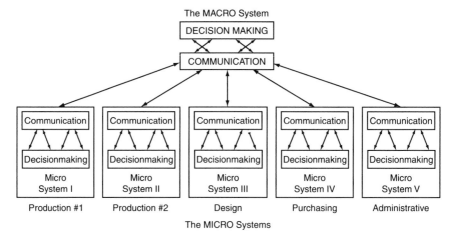

Figure 17-6 The macro and micro monitoring/measurement systems.

When the component parts (the processes at the microlevel) are working well, the whole organization is working well. To create a useful, effective measuring system, the organization-level (or macrolevel) issues have to be surfaced. For each issue there are a series of macrolevel measures to be established. Then, the underlying processes need to be identified and appropriate measures established for each of them. This sequence is similar to the process used in Policy Deployment (also known as Hoshin Planning).

The macrolevel measurement development (see Figure 17-7) starts with a review of (or establishing) the organization's mission and vision statements. These two instruments guide the discussions on the organization's product/service offerings. Once the near- and long-term offerings are established, the customers for these offerings are identified. For an organization that seeks to be customer driven and wants to get as much response as possible in the marketplace for the resources invested, it is essential that an ongoing process of customer information gathering be initiated. The environment, marketplace, different customer categories and customer groups within categories are constantly changing. For this reason it is necessary to acquire knowledge of customers' expectations and then to create or adjust measures based on these dynamic customer inputs.

For different groups of customers, the services they will expect and under what conditions should be determined. Will they be looking for

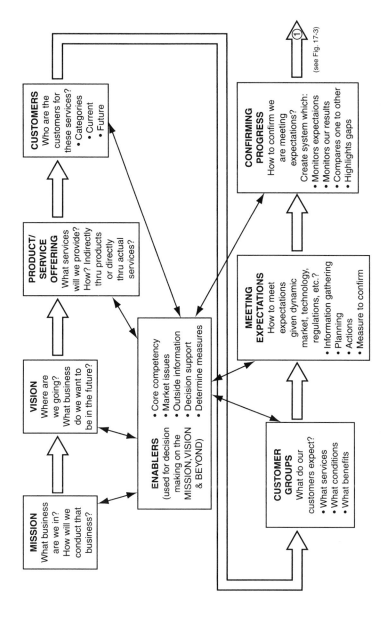

Figure 17-7 Measurement system development process steps (macrolevel).

MISSION
What business are we in? How will we conduct that business?

VISION
Where are we going? What business do we want to be in the future?

PRODUCT/ SERVICE OFFERING
What services will we provide? How? Indirectly thru products or directly thru actual services?

CUSTOMERS
Who are the customers for these services?
• Categories
  • Current
  • Future

ENABLERS
(used for decision making on the MISSION,VISION & BEYOND)
• Core competency
• Market issues
• Outside information
• Decision support
• Determine measures

CUSTOMER GROUPS
What do our customers expect?
• What services
• What conditions
• What benefits

MEETING EXPECTATIONS
How to meet expectations given dynamic market, technology, regulations, etc.?
• Information gathering
• Planning
• Actions
• Measure to confirm

CONFIRMING PROGRESS
How to confirm we are meeting expectations? Create system which:
• Monitors expectations
• Monitors our results
• Compares one to other
• Highlights gaps

① (see Fig. 17-3)

partnerships or arms-length transactions? Will what is sold be a product or access to a service? (Will the customer want to own the truck or lease it? Or would he or she be more satisfied to have the freight transportation services provided by a trucking service? Perhaps with trucks that have the customer's name emblazoned on the side!) Using a variety of sources of customer inputs and expectations, an organization should create measures that specifically respond to customers' expectations as well as a system to assure that these measures are meeting customer expectations.

The last and most important step is to create a system that compares customer expectations for each customer group with an organization's results. The difficulty is to form similar measures that can easily be compared. Measures of expectations in customer "jargon" and measures of results in technical language are not often comparable. In situations like this, it is valuable to use the Voice of the Customer Table (VOCT). The VOCT assists in translating customer jargon to technical language. Once these measures have been established, they can be compared. (A word of caution: Measurements that assess levels of customer satisfaction and measurements that assess levels of organizational effectiveness are neither trivial to create nor easy to compare.) For an organization to maximize its benefits, a system must be designed to monitor the gaps and then communicate them throughout the organization in a timely fashion. In this way difficulties can be anticipated, avoided or mitigated and opportunities exploited.

IV

MANAGING AND MONITORING QFD

18

MANAGEMENT'S INFLUENCE ON QFD

Done well, a QFD project coordinates resources drawn from throughout the organization, adds expertise on a continuing or ad hoc basis from key suppliers and customers and comes up with a product or service that is *beyond* state-of-the-art. Furthermore, it is introduced to the market earlier than anyone ever thought possible, at the onset of the project. For these expected outcomes to become a reality, it is necessary for the resource controllers (primarily those who have "lent" their most knowledgeable people to the project) and stakeholders (those who are going to have "mud" on their face if the project fails) to not only be aware of the project's progress but also to be informed on a timely fashion of any impediments.

Many of the challenges for a specific QFD project may be unknown until encountered. However, as the following SWOC (Strengths, Weaknesses, Opportunities, Constraints) analysis shows, some problems are common to all QFD projects. To increase both the chances and level of success, management must be aware of and address these problems. By providing timely support, management can have a tremendous impact on the morale and effectiveness of the QFD Action Team.

SWOC ANALYSIS ON QFD

Strengths

1. Structured (more effective, less waste of resources)
2. Planned up front (fewer risks, greater knowledge gathered and greater likelihood of accomplishment)

3. Overall cost is less per development program due to
   a. Fewer changes to be made later because of the extra work done in the conceptual stage
   b. A greater proportion of the changes made occurring in earlier stages when they are easier and cheaper to make
   c. Reduced development time—achieved as a result of increased efficiency, greater information flow and more sharing of information in a structured format
4. Return is greater because
   a. Market entry is earlier
   b. Market share is earned more easily
5. Life-cycle costs are lowered because
   a. Product's entire cycle is dealt with as an integrated whole—from concept to production to customer use to disposal/recycling, thus resulting in a better design
   b. Better designed product or service resulting in
      i. Greater customer satisfaction
      ii. Fewer returns (on product), fewer complaints (on service and product)
      iii. Fewer warranty claims
      iv. Lower service parts inventory needed
      v. Less maintenance and service needed

Weaknesses

1. Needs long-term (3–6 months plus) commitment from many different (conflicting?) segments of the organization at both a high level and a low level
2. Is a new approach that
   a. May not have a champion
   b. Requires additional up-front work (compared to other design approaches)
   c. Requires the organization to address many QFD tasks it has never done before and to integrate the results
   d. Requires teamwork of a group of (often) near strangers who may not (often do not) have team-building skills or experience to be effective
   e. Is difficult to institutionalize because it is a one-time event in the case of a QFD product or service development project (although some persons on the first project team may serve on later project teams and then the QFD knowledge is deployed)

Opportunities

1. To reduce development costs in the near term
2. To reduce manufacturing and distribution costs in the midterm
3. To reduce life-cycle costs in the long term
4. To be more responsive to the market with a more flexible product or service
5. To deliver higher quality (addressing customers' needs with fewer gaps and missed opportunities for greater competitive impact) in shorter time (from concept to fully ramped production) at lower cost (initial and lifetime).

Constraints

1. Since a QFD team runs across several "jurisdictions," it can easily be made ineffective by
   a. Political moves
   b. Withholding or reducing resources
   c. Management not making adequate provision for QFD team members to be relieved of their day-to-day ongoing responsibilities (a de facto reduction in team resources)
2. Because QFD is a sequential process, a blocked step cannot easily be worked around

In today's competitive world, management wants and needs the benefits of QFD. The main vehicle for attaining these benefits is the QFD Action Team. So, understanding the Action Team's tasks and potential problems is important to getting the desired and required results and benefits. Senior and middle managers who are aware of the difficulties of doing QFD can positively impact both the process and results. To address the weaknesses outlined in the SWOC analysis above, management should have an overview session (see 2 in the next section, under Typical QFD Project).

From past experience, the three most important issues to be covered with managers are:

First Issue. Visible leadership at the senior management level is an absolute necessity. All functions (and department managers) must understand the "strategicness" of the QFD effort and the necessity for cooperation between the various functions. To accomplish the common goal of a superior design in a shorter time frame, the senior management team must understand and address this issue.

Second Issue. When team members are withdrawn or switched during the project, there is a corresponding reduction in the effectiveness of the Action Team and lowered chances for success of the QFD process. Department managers must understand and formally address this issue.

Third Issue. Acknowledge the need for timely action when an impediment exists to the QFD process. Both senior and middle managers must accept their responsibilities to address, eliminate or mitigate this barrier.

Senior management can address these issues and show leadership by

1. Attending the Overview Session for Management (4 hour meeting) (see 3 under Typical QFD Project)
2. Having at least one senior manager address the first Action Team meeting (the "kick-off" meeting) on (a) why QFD is strategic, (b) that all of management is committed to QFD and will support the Action Team's activities and (c) senior management will assist managers to solve problems and overcome impediments, where necessary (see 4 under Typical QFD Project)
3. Attending the Process Review (De-brief) Session (see 6 under Typical QFD Project)
4. In general, receiving and acting on communications from the Action Team (see Figure 18-1) or affected managers

Middle managers can address these issues and show support for the QFD Action Team as follows:

1. By carefully considering which subordinates to assign to the Action Team, from both knowledge and continuity perspectives. A typical Action Team only has one person from each function representing it on the team. To the extent possible, that person should be very knowledgeable about the product/service being designed (and the related processes) *and* should be available for the *entire* time of the project. [Although the QFD process creates a very good "trail" of information (what was done, what decisions were made and the basis for those decisions), it still takes a long time for a replacement individual to get "up to speed" and to be a good representative of their function as well as a contributing member of the Action Team. Loss of one team member usually cuts the Action Team's effectiveness by a third (even if that person

Team member or resource	Knowledge base	Stakeholder and tittle	Comm. mode	Content. of comm.	Frequency	Responsibility
1. Sally Pine	Customer service	Steve Black, Mgr. C.S.	Phone call	Problems C.S. can assist on	Every mtg.	Sally Pine
2. Ron Dunn, team leader	Marketing	Pat O'Brien, VP Mkting	Company e-mail	Activities summary	Every week	Ron Dunn, team ldr.
3. John Sterling	Design eng.	Alex Smith, VP Mkting	Company Co mail	Progress to plan	Every 2 weeks	Fred Park, facilitator
4. Tim Anders	Mfg. eng.	Ben Arnold, VP Ops.	Company E-mail	Memo on problems	Every qtr.	Ron Dunn, team ldr.
5. Pat Bell	Quality assurance	Alan Span, Mgr. Q.A.	Phone call	Meeting summary	Every mtg.	Fred Park, facilitator
6. Sam Johnson	Field Service	Jo Welch Service Mgr.	Company E-mail	Progress summary	Every week	Sam Johnson
7. Ann Williams, elec. engr	Supplier: AGP Tronics	Arnie Jones VP Engr. AGP	Faxed memo	Future opps & current progress	Every month	Ron Dunn, team ldr.
8. Shel Retlin, marketing	Customer: King Elec.	Jon King, III, VP, King Elec.	Internet E-mail	Meeting Minutes & Comments	Every mtg.	Pat O'Brien
9. Fred Park, facilitator	QDF Process & team building	Alice Rose, Dir. H.R.	Write-up Co. mail	Summary, problems & progress	Every month	Fred Park, facilitator
10. QFD process, action team	Re-engineered model R10ce	Bob Randel, Presiodent.	Company E-mail	Summary, problems & progress	Every 2 wks	Pat O'Brien, Alex Smith

Figure 18-1 Stakeholder Communication Plan: example.

is replaced), loss of two members will reduce effectiveness by two-thirds and loss of three team members usually means the end of that Action Team. *Note:* these comments only apply to the full-time, core Action Team members, such as 1–6 of the team member column in Figure 18-1.]

2. By receiving and acting on communications from the Action Team (see Figure 18-1), especially those items that are an impediment to the team's progress.

3. By attending the management meetings: Overview Session for Management (4-hour meeting) (see 3 under Typical QFD Project) and Process Review (De-brief) Session (see 6 under Typical QFD Project).

To be successful in implementing the QFD approach to design, leadership, vision and, most of all, commitment to the process at both the senior and middle management levels are required. If the Action Team senses a lack of commitment, it will have a strong negative impact and jeopardize success. It is essential that the scope and vision of the QFD project be well defined. Without that there will be project "scope creep." In the QFD process, especially during the translation of the Voice of the Customer, there will be many more discoveries made than usual about how the customer uses the product or service and what the customer wants in a product/service. Often, there will be a strong tendency to want to do more than was originally intended when the project was defined. During the competitive analysis, more research is often done on more competitors and about more features and functions than ever before.

With all this extra information, there will be a strong temptation to want to include more features and functions in the QFD project's product or service. The Action Team and management must resist this temptation. The focus must be on the *benefits* derived from the Voice of the Customer analysis. Scope creep occurs when the Action Team tries to include in the design all of the features and functions in all of the competitor's products/services on the market plus additional ones. In other words, concentrate on the *benefits* that customers tell you they want, not on the *features and functions* of all the competitive products or services already in the market.

This commentary should not be interpreted in any way to restrict the Action Team's efforts to include Excitement Quality elements in the product/service design. It is intended to raise the awareness that more features and functions and/or copying other designs' features and functions does not necessarily make the product/service better or more desirable in the marketplace. We are all familiar with consumer electronics products that contain more "bells and whistles" than anyone typically uses, even to the point that the average consumer is confused, frustrated, overwhelmed (pick one, pick two!!) by the variety of options. When management reviews the Action Team's progress, there must be a constant awareness of the potential for scope creep and a continued refocusing of the team's activities.

One of the most effective ways to keep managers engaged in the QFD process and to reduce chances that essential team members will be withdrawn (Issue Two) or that a problem or impediment will be ignored until it has sapped the Action Team's resources or reduced the team's momentum (Issue Three) is to establish and use the Action Team Stakeholder Communication Plan shown in Figure 18-1. The plan takes into account that different people are comfortable with different levels of information (content and frequency) and often have different opinions on how they want to communicate and from whom they want to hear. The Communication Plan takes all of that into account. Experience has shown that the Communication Plan is a preventive measure, essential for the success of a project such as QFD. The nature of projects is that they are easy to start but difficult to keep going and keep on track (sustain and guide). The Communications Plan is a tremendous help in overcoming those problems.

The Communications Plan applied to the QFD project is important to maximize the utilization of the project's resources and, hence, its resulting success. Some organizations use similar plans throughout the organization at the tactical level so as to increase their effective and timely use of resources. The ultimate, however, is a planning approach that involves all the functions of the organization and all of their tasks for at least 12 months. This approach is called Policy Deployment, also referred to as Hoshin Kanri, Hoshin Planning or Management by Policy (MBP). Policy Deployment is an alternative approach to Management by Objectives (MBO) and generates its power from the manner in which it is deployed. At the outset, the senior staff sets certain goals for the organization based on its mission and vision and shares them with the subordinate functional areas. Then each functional area takes the organizational goals and determines what they must do to support the higher level goals. At the same time they determine the resources necessary for their functional area to attain those goals.

The functional area then returns their statement of goals and the resources needed to accomplish those goals to the next higher level. If the higher level is willing to allocate those resources, then a contract is struck: this functional area has these resources assigned to accomplish these goals. Sometimes, however, the higher level hierarchy is unwilling or unable to allocate the complete set of resources requested. When that happens, a series of events called "catch ball" occurs in which the two levels go back and forth trying to strike a balance between what resources can be allocated and what accomplishments need to be committed to.

Once agreement is reached between each functional area and its higher level, then within each functional area, going down into the organization level by level, catch ball is played until, in the end, the high-level goals have been deployed to the lowest level of the organization. At this point, everyone in the organization has agreed on what their goals are and the resources they will have available to reach those goals. For every area, there will be different targets, or *hoshins* (*hoshin* means bright shiny object or target in Japanese), which the personnel in that area will monitor closely. This is similar to the QFD project's Communications Plan, where information is constantly monitored and fed to decision makers to assure the project's goals are being met, but on a much larger scale. It affects everyone in the organization, not just the QFD Action Team and the impacted managers.

The Communications Plan does not outline the entire Action Team–management interface. Besides receiving and reviewing the communications from the Action Team, management members will be involved in some additional activities. Based on individual circumstances, different firms will begin a QFD project at different stages. The following is an example of a typical QFD project with details of management involvement. (This overview assumes that the QFD methodology is not initially well known, either to management or to the Action Team members, and that training on the QFD technique will need to be provided.)

TYPICAL QFD PROJECT

1. *Scoping and Operations Review (1–2 days).* This step is done by the project (team) leader and trainer/consultant (internal or external person knowledgeable about the QFD process) with the purpose of understanding the organization and product or service that will be the subject of the QFD project. Data and nomenclature are collected that are relevant to this QFD project's product or service and that can be used in the QFD training sessions. Using the organization's own information will make the training more effective for the participants. As part of the information-gathering tasks, some managers may be interviewed to get their ideas on a good example to use and sources of pertinent data/information.

2. *Overview Session for Management (4 hours).* It is vital that managers who will have subordinates who are part of the QFD Action Team as well as the senior executive team attend this overview session. Other managers should also be encouraged to attend. The goal is to inform managers what QFD is, what it can and cannot do, what the benefits are, what it requires of them (cooperation between functions) and what are the "correct" questions to ask at which points so as to manage the process and, hopefully, create some champions for QFD within the organization. The overview session also covers the process of choosing team members and the importance of continuity and knowledge in that process. The Communications Plan between the QFD Action Team and the managers/stakeholders is created during this session (except for the listing of some of the team members). All impacted managers should be involved in the 4-hour meeting.

3. *Training Session for Full- and Part-Time QFD Action Team Members (3–4 days).* All persons who will work on the QFD Action Team will go through the same training. Part-time or ad hoc members such as those from key suppliers or customers or from corporate staff will receive the same training as the full-time members. The goal is to inform all team members and interested managers about the elements of the QFD process and what to expect as they go through the process, both from an analytical/technical standpoint as well as from a behavioral/team-building standpoint. At least one example will be worked through based on the information gathered during the scoping and operations review. Interested managers are invited to participate in this and the next training session. All interested managers should be involved for the full 3–4 days.

4. *Training for QFD Facilitators and Internal QFD Consultants (3–4 days in 3 above plus 2 days).* The training is for both QFD facilitators working with the initial QFD projects and persons responsible for propagating the QFD technique within the organization. The goal is to have a cadre of personnel familiar with the technique who can perpetuate it after the pilot consultant/trainer facilitated project is completed. Managers interested in championing the QFD technique are strongly encouraged to attend. All managers interested in being champions should attend for the 3–4 days of session 3 plus 2 days.

5. *Facilitation of QFD Action Teams (typically 1 day each week for 3–6 months).* The goal is to take the QFD team members, facilitators and propagators through a complete program from concept to product/ service design. Also, to set up a "postmortem" review structure for future updating of the matrices involved and setting up how to conduct feedback to the correct organizational elements to facilitate continuous improvement in the design of products and services. Managers are usually not present at most of the regular QFD Action Team meetings. However, at the completion of the QFD pilot, it is essential that the affected managers (document control, design engineering and manufacturing engineering) attend the meeting(s) for creating the ongoing review structure for updating and feedback. Document control, design engineering and manufacturing engineering managers should attend approximately four meetings of 2 hours each to define the updating and feedback structure for QFD-related information.

6. *Process Review (Debrief) Session with Management (4 hours).* The goal is to go over with the management team the just completed QFD project, point out the lessons learned and share the outcomes (both in terms of the new product/service designed as well as the now-trained

QFD Action Team members and their team-building and design skills). Side benefits of this review session are increased awareness of QFD and, hopefully, reinforced convictions of current QFD champions as well as the creation of additional champions. All managers should attend the 4-hour debriefing meeting.

19

MANAGEMENT'S NINE ACTIONS FOR QFD ANALYSIS

Nine actions that are collectively management's responsibilities relative to QFD Analysis Teams have been sanctioned in their organizations:

1. Study and fully understand the QFD process from beginning to end.
2. Identify and prioritize projects requiring QFD analysis.
3. Define the purpose and plan resource allocation for QFD Projects.
4. Develop the Goals, Organizational Objectives, Scope and Expectations (GOOSE) for each QFD Project (Figure 19-1).
5. Appoint a QFD Project program manager.
6. Appoint a QFD facilitator and a cross-functional QFD Analysis Team.
7. Provide a charter to the QFD teams.
8. Enable, empower and encourage QFD teams.
9. Provide commitment and involvement to QFD teams by using a management MAP (Monitor, Audit, Probe) process.

These management responsibilities are explored in detail in the following pages. In addition, we have included several new forms for management's use in their execution of the QFD review process.

ACTION 1: UNDERSTAND THE QFD PROCESS

As with any management tool, procedure or process, QFD is best understood if it is studied and discussed with experienced Subject

Matter Experts (SMEs) prior to involvement. While understanding of the QFD process and procedures is important, experience in its application is a major time and cost saver for QFD neophytes.

Management would be well advised to listen to and work with external or internal QFD consultants prior to initiating any QFD Projects within their organizations. Additionally, there are a number of publications available for self-study by management. These include several texts (see the Bibliography) and the proceedings of the annual QFD Symposiums at Novi, Michigan (see Appendix B).

QFD is concerned with the conceptualization, design, development, production, assembly, packaging, shipping and functioning of new and enhanced products and services that are responsive to the Voice of the Customer. However, other voices require attention by management: those of the engineer, the process, all the support functions and society as a whole.

QFD is not a panacea that can be expected to resolve all of management's concerns and solve all of an organization's problems. Fortunately, QFD smoothly interfaces with a broad variety of Total Quality Management tools, for example, Concurrent Engineering (CE), Cost of Quality (COQ), Cycle Time Management (CTM), Design of Experiments (DOE), Input/Output(I/O) analysis, Nominal Group Technique (NGT), Statistical Process Control (SPC), and Value Engineering (VE). These important relationships are.

QFD and Concurrent Engineering (CE)

- QFD emphasizes early involvement of all company functions.
- QFD is a natural adjunct to CE.
- QFD acts as a framework and catalyst for CE by
    - Providing a systematic methodology for integrating interdisciplinary inputs
    - Facilitating better interdepartmental communications

QFD and Cost of Quality (COQ)

- QFD focuses on creating least cost products to meet customer needs.
- COQ supports QFD by highlighting the cost of not doing it "right the first time."
- COQ identifies cost of conformance (prevention and appraisal) and cost of nonconformance (internal and external failures).

- COQ recognizes the most significant costs and their causes.
- COQ databases serve as important sources of information to drive QFD efforts.

QFD and Cycle Time Management

- QFD is concerned with providing improved designs in less time.
- CTM provides QFD Project teams with a methodology needed to
  - Clearly define process purpose and objectives.
  - Identify process start and end points.
  - Highlight value-added and non-value-added activities.
  - Identify tasks' times, choke points and duplicative operations.

QFD and Design of Experiments (DOE)

- The QFD process helps the project team select those factors that need further support.
- DOE supports QFD through its capability to
  - Identify controllable and uncontrollable factors.
  - Efficiently and effectively conduct experiments to discover interactions and effects between factors.

QFD and Input/Out-Put (I/O) Analysis

I/O analysis supports QFD because it provides

- Clarification of roles and responsibilities
- Resolution of roles and responsibilities
- Elimination of duplication
- Opening of the lines of communication

QFD and Nominal Group Technique (NGT)

NGT supports QFD because it provides

- Generation and presentation of ideas
- Clarification of logic and data analysis
- Development of action plans

QFD and Statistical Process Control (SPC)

- QFD provides insight into what needs to be controlled to ensure a stable, design capable process.
- SPC ensures that gains achieved through the use of QFD are not lost.
- SPC can help to
  - Identify areas for performance improvement.
  - Discover common and special-cause problem sources.
  - Prioritize problems and identify causes.

QFD and Value Engineering (VE)

QFD supports VE by providing a mechanism to

- Identify required functions.
- Analyze current configurations.
- Identify costs associated with each part.
- Determine the cost to perform required functions.
- Develop alternative configurations.

ACTION 2: IDENTIFICATION AND PRIORITIZATION

At any given point in time most organizations have a multitude of projects that should be analyzed using QFD. A facilitated brainstorming session composed of management drawn from across the broad spectrum of organizational functions will generate a comprehensive listing of potential QFD Projects.

Application of the prioritization matrix, one of the seven management and planning tools, will result in a prioritization of the project listing. We recommend that management allocate its scarce human and material resources to this prioritized list from the top down.

Prioritization is required since most organizations do not have all the resources necessary or available to do everything at the same time. There are a number of prioritization processes, ranging from voting techniques to a more rigorous pairwise comparison process.

The pairwise comparison process is the recommended approach since it is more rigorous than other prioritization methods. In the pairwise comparison method each project is compared to all the other projects in a two-at-a-time fashion. The Prioritization Matrix (see Brassard,

1989) Plus is a pairwise comparison process that uses the following scale to develop a weighting of each project's importance in accomplishing the organization's stated goals:

Weight	Relative Importance
1	The two projects are of equal importance in accomplishing the organization's goals.
5	One project is significantly more important than another project in accomplishing the organization's goals.
10	One project is exceedingly more important than another project in accomplishing the organization's goals.
1/5	One project is significantly less important than another project in accomplishing the organization's goals.
1/10	One project is exceedingly less important than another project in accomplishing the organization's goals.

A total score is computed for each issue project. Then the projects' scores are rank ordered to assist the management team members in deciding on the most important projects to pursue.

This process develops a team consensus about the path to follow and instills a sense of ownership in the paths chosen. This ownership creates the will or spirit in the team to follow through.

ACTION 3: PROJECT PURPOSE AND RESOURCE ALLOCATION

Each QFD Project must have a specific, well-stated purpose. Creation of this statement of purpose is as much for the management that defines it as for the QFD Analysis Team that will ultimately study it to better understand their team charters. A statement of purpose must precede any other statement relative to the creation of a QFD team. Figure 19-1 shows a typical QFD Analysis Team Charter Document. Some of the sections in Figure 19-1 will be discussed in more detail in Action 4.

Management has a finite limit of scarce resources available to be allocated to an infinite number of projects, QFD and otherwise. How much of each resource (human, time, dollars, space, equipment, etc.) will be allocated to any QFD Project? Figure 19-2 is a matrix analysis form designed to help management plan this allocation of resources to the various QFD Projects selected for study. It is essential that the appropriate resource level be allocated to each team to accomplish

G.O.O.S.E.

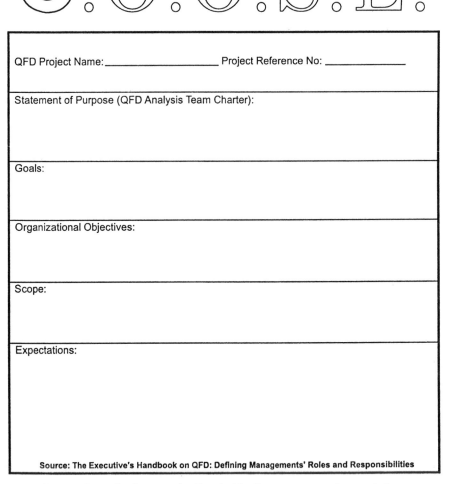

QFD Project Name:_____ Project Reference No: _____

Statement of Purpose (QFD Analysis Team Charter):

Goals:

Organizational Objectives:

Scope:

Expectations:

Source: The Executive's Handbook on QFD: Defining Managements' Roles and Responsibilities

Figure 19-1   Goals, organizational objectives, scope and expectations.

their mission. The lack of the appropriate level of resources causes a QFD Analysis Team discouragement, frustration and stress in not being able to achieve their objective(s) due to external factors that are not within their control.

The resource allocation matrix is a systematic way to compare potential QFD Projects, one to another, so that scarce organizational resources can be allocated effectively. The left-hand side of the matrix

QFD Project Name	Resources Required/Allocated					
	Human	Time	Dollars	Space	Equip.	Other
PF3	30 People	4 Months	1000K	10,000 sq. ft.	Furniture/ computers	
Total Required	30 People	4 Months	1000K	10,000 sq. ft.	Furniture/ computers	
Total Allocated						

Source: The Executive's Handbook on QFD: Defining Managements' Roles and Responsibilities

Figure 19-2   QFD Resource Allocation Matrix.

will contain a listing of the potential QFD Projects. The top of the chart shows the resources most commonly required. Each listed project should have a listed estimate of the quantity of each resource it will require. When all projects have their resource needs entered, a total for each can be tallied. Management can then compare the total required resources to the availability of resources. Choices must be made by management as to which projects will be approved and which ones will be delayed. The allocated portion of the matrix is filled in as the project progresses and resources are allocated.

Recognizing that management's responsibilities relative to preparing for a QFD Project are both broad and deep, a detailed checklist has

1. Determine nature of the QFD Project in terms of outcomes, expectations, and benefits:

   ☐ Describe process or program to be evaluated.
   ☐ Why was this project selected?
   ☐ List goals and desired outcomes of this QFD Project.
   ☐ Anticipate how this project will be improved as a result of QFD.

2. Determine QFD team membership and logistics:

   ☐ Determine selection criteria.
   ☐ Select core team membership.
   ☐ Select team leader.
   ☐ Identify QFD facilitator.
   ☐ Select other cross-functional members for possible support.
   ☐ Determine team logistics: location, meeting schedule, when, where, how long, etc.
   ☐ Decide frequency of joint reviews with both management guidance team and QFD team.

3. Determine QFD team boundaries and support:

   ☐ Designate start and finish dates of QFD Project.
   ☐ Identify project resources, funding support, budget limitations, etc.
   ☐ Specify QFD team's authority to request or specify other needed resources.
   ☐ Determine QFD team training schedule.
   ☐ Identify QFD trainers.
   ☐ Identify subject matter experts (SMEs).
   ☐ Describe how much times each team member is to allocate to this QFD Project.

Figure 19-3 QFD Project preparation management checklist.

been developed. Figure 19-3 provides a multiphase approach for management's use.

ACTION 4: THE GOOSE

The GOOSE are project parameters that must be developed for each QFD Project before any personnel are identified for assignment to the project. An example is shown in Figure 19-4. First, generic goals are addressed, for example, "reduce product development time." Second, more specific measurable objectives are derived from the generic goals, for example, "reduce product development time by 40% from the current average of 240 days." Third, the scope of the project must be defined, for example, "the project will address W, X and Y but not

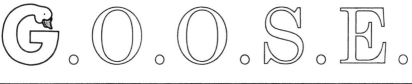

QFD Project Name:_____ Project Reference No: _____
Statement of Purpose (QFD Analysis Team Charter):
Goals: **Reduce product development time.**
Organizational Objectives: **Reduce product development time by 40% from the current average of 240 days.**
Scope: **The project will address W, X, Y, but not Z.**
Expectations: **A justification for adoption of a new concept and/or technology will result.**   Source: The Executive's Handbook on QFD: Defining Managements' Roles and Responsibilities

Figure 19-4   Goals, organizational objectives, scope and expectations (example).

Z." Fourth and finally, management must declare its expectations for the project, for example, "a justification for adoption of a new concept and/or technology will result."

A Tree Diagram (see ASI, 1992) is a useful tool for developing a generic goal into its specific subgoals. In increasing detail, it systematically maps the full range of tasks that need to be accomplished to achieve a general goal and related subgoals. This tool helps keep a team focused strictly on the details related to the goal being considered.

A Tree Diagram is developed from left to right asking the following question at each level: "What needs to happen for this to be accomplished?" This question helps to focus the development of the Tree Diagram in an increasingly more logical level of detail. In this process there is a tendency to make large leaps in the detailing process and not have the logical connecting flows. Then, a second series of questions are asked as the Tree Diagram is checked for completeness. This check proceeds from right to left, from the details to the generic goal. The question we ask from right to left is "If these items happen, will the next level occur?" If we made large leaps in detail, we would see the disconnect very quickly. The Tree Diagram approach assists in filling out the GOOSE, as shown in Figure 19-4.

ACTION 5: PROJECT PROGRAM MANAGER

Every QFD Project requires a project program manager who will interface with management and provide the leadership necessary to ensure successful completion of the project. Appointment of the right person to fill this position is a critical juncture in the chronology of the QFD Project. Such a person must simultaneously be able to effectively respond to questions and concerns posed by management as well as to smoothly lead the QFD team from phase to phase. The project program manager must lead the team with an efficient use of resources and the effective coordination of team members.

This project program manager will be responsible for steering the efforts of the QFD Analysis Team and integrating the results of the study into the overall product (or service) development process. Under no circumstances should this responsibility be delegated by the project program manager.

The project program manager also has responsibility for creating a Communications Plan that responds to and suits the unique requirements of each stakeholder and resource controller who is impacted by the QFD project. In the case of communications, some reports, written or oral, may be delegated to QFD Analysis Team members for them to carry out, often with the communications being directed to the functional area that the team member represents.

The following is a list of the various roles and responsibilities a project program manager of a QFD team will be required to perform:

Roles

- Coach
- Facilitator
- Planner
- Referee
- Trainer
- Mentor
- Arbitrator

Responsibilities

- Coordinate QFD team activities
- Focus QFD team on concept being developed
- Make decisions using consensus techniques
- Ensure that team functions efficiently and effectively
- Track and measure progress
- Obtain needed resources
- Be a technical resource
- Reward and recognize team accomplishments
- Motivate team to reduce its cycle time of product (service) development
- Present team's progress
- Act as liaison between team and management
- Those determined by your QFD team's unique needs and opportunities

This listing can be used by the management team as a guide in evaluating suitability of prospective project program managers.

The project program manager must interface with management, R&D, engineering, marketing and sales. Figure 19-5 shows the roles and responsibilities of these functions as well as a summary of the project program manager's job.

The following excerpt from the General Motors Quality Network Reference Guide, 1990 shows an example of detailed group objectives and responsibilities for QFD:

Responsible	Strategy	Feasibility
Senior management	• Prioritize opportunities • Approve Concept Development Plan • Approve first pass of business plan • Approve concept feasibility budget • Select and assign program manager	• Approve Product Development Plan • Approve final pass of business plan • Approve development budget • Prioritize resource expenditures
R&D Plus engineering	• Assess technology feasibility • Develop product development plan • Assess and prioritize technology • Determine strategic fit of technology	• Develop product platform and architecture • Develop system specifications • Complete feasibility analysis
Marketing & sales	• Identify growth markets • Identify market need • Conduct market research studies	• Develop market plan • Conduct customer surveys • Conduct field studies • Develop distribution strategies
Project program manager	• Project program manager is appointed after strategy is developed and need for a program is defined	• Assemble cross-functional team • Manage product development plan • Manage and coach team • Complete product delivery plan • Track and report on progress • Set goals for team • Set priorities • Coordinate internal communication • Scope manufacturing plan • Set performance goals

Figure 19-5 Roles and responsibilities in developing products and services.

GROUP OBJECTIVES AND RESPONSIBILITIES

Corporate

- Chairman, President, Executive Vice Presidents, International Union Officials

Objective

- Support and commitment for the implementation of the QFD methodology throughout the Corporation through leadership.

Responsibilities

- Understand QFD as one ongoing process which would enhance the integration of Corporate functional activities throughout the Four Phase Process.
- Require the integration of QFD within the corporation by promoting the implementation of the Quality Network QFD Action Strategy.

Group/Division Leadership

- General Managers, Vice Presidents, Union Leaders, Business Unit Directors

Objective

- Ensure the integration of QFD in all future product programs consistent with the Four Phase Process and other Quality Network strategies. Be consistent in leadership style and provide a supportive environment that will encourage the positive participation to all GM employees.

Responsibilities

- Promote QFD as an ongoing process rather than an event.
- Be patient enough to realize its benefit.
- Invest resources early in product programs to carry the Voice of the Customer into our products and processes.
- Ask for the results obtained from the QFD review.
- Incorporate QFD as a part of the business plan for product and process development.

- Assure integration and coordination of QFD activities among vehicle divisions, platforms and suppliers.
- Ensure that QFD information drives product and process decisions.

Platform/Staff Management

- Program Managers, Platform Engineering Directors, Vehicle System Engineer, Vehicle Subsystem Engineer, Component System Engineer, Component Subsystem Engineer, Staff Product/Process Development Engineer

Objective

- Lead QFD process and create an atmosphere of teamwork to enable deployment.

Responsibilities

- Understand the QFD methodology and integrate into their organization.
- Acquire QFD training and maintain skill base in QFD methodology.
- Assume responsibility for the integrity of system/subsystem teams throughout product and process development.
- Organize and conduct QFD project reviews.
- Assess and satisfy QFD training needs.
- Assume completion of the QFD-AST Implementation Plan.

QFD Team

- Cross-Functional Teams for: Vehicle, Subsystems, Components, Staff and Product/Process Development Engineers

Objective

- Execute the QFD process from planning through production.

Responsibilities

- Acquire QFD training and maintain skill base in QFD methodology.
- Obtain, analyze and record information required.

- Resolve any conflicts in alternative product and process technology and measurements (trade-offs).
- Offer suggestions for continuous improvements.
- Communicate results and recommendations at review.

Internal Customers and All Suppliers

- All Affected Employees: Salary and Union

Objective

- Provide/develop products, processes or services which respond to the Voice of the Customer through direct or indirect involvement in the QFD process.

Responsibilities

- Participate in and/or lead QFD studies, as appropriate.
- Provide information consistent with requests.
- Understand the GM QFD process as it relates to them.
- Implement the output of QFD.

Education and Reference Material

QFD is a strategic issue for General Motors. Consistent application of the QFD process requires uniform education and training of the subject matter. To achieve this uniformity, the development and delivery of QFD training requires coordination. In this regard, the QFD Action Strategy Team is developing the GM training strategy for QFD. In support of these education and training initiatives, in GM's publication there is a list of available resource materials in this section. These resource materials plus others are listed in the Bibliography.

ACTION 6: PROJECT FACILITATOR AND CROSS-FUNCTIONAL ANALYSIS TEAM MEMBERS

The newly appointed QFD Project program manager, along with the management team that appointed him or her, must select a QFD Project facilitator as well as the other members of the QFD team. The facilitator's function is to

- Ensure the team performs its assignment smoothly.
- Coordinate and set up the meetings.
- Help the project manager overcome barriers.
- Prevent team discouragement and frustration due to lack of progress.
- Help keep the project on track.

In many ways the facilitator is both a gatekeeper and timekeeper, that is, a QFD parliamentarian.

The QFD Analysis Team members, like their management counterparts, should be drawn from across a broad spectrum of functions applicable to this QFD Project. This cross-functional team composition will reduce the likelihood of omission or lack of interfunctional coordination and other bureaucratic oversights.

QFD team members should not be lower level staff personnel who have little or no knowledge of existing organizational commitments. They must be

- Knowledgeable, either technically or in the customer/marketing aspects of the QFD
- Ready to articulate their function's capability to perform and capacity to act, that is, be dependable spokespersons for the functions that they represent
- Able to make decisions that do not have to be reviewed

If QFD team members are unable to perform at this level, then considerable time will be wasted and the other team members will be frustrated in their efforts to proceed in a timely and effective manner.

ACTION 7: QFD TEAM CHARTER

Integration of the GOOSE produces the QFD team charter. At the first meeting with the QFD team facilitator and the other QFD team members, the QFD Project program manager should distribute the team's Statement of Purpose (defined by management in Action 3). The team needs to discuss the Statement of Purpose, clarify what it means to them and use it as a basis for creating/refining the QFD Team Charter and the QFD Project's name. The charter statement should read in a manner that allows the team to also create indicators of success.

During the initial session the QFD team members should be encouraged to review and challenge the entries in the GOOSE form as well as the linkage and alignment of the QFD Project to the organization's goals. The team members' questions can lead to modifications of the initial entries as well as the raising of issues that were not previously or adequately considered. Table 19-1 is a listing of 10 concerns that teams can use as a QFD Project Selection Checklist to assist in this reviewing task.

Prior to addressing the 10 concerns, the team might want to consider some or all of the following points for discussion.

1. (a) is this a brand new product or an improvement on a product/ service that was already being delivered? [To maximize your chances of success, you do NOT try to learn a new methodology (QFD) while designing an entirely new product or service.]

(b) Is the initial product/service being improved closely related to others and can the knowledge gained from the initial QFD effort be easily replicated in other product/service lines [the more replication possible, the greater the return on the first (pilot) QFD Project].

Table 19-1   Project Selection Checklist

Yes	No	
⎯⎯	⎯⎯	1. The QFD Project is related to key business issues.
⎯⎯	⎯⎯	2. The product/process targeted for improvement has direct impact on the company's external customers.
⎯⎯	⎯⎯	3. The QFD process or program has visibility throughout the company.
⎯⎯	⎯⎯	4. All of the managers concerned with this process or program, at all levels of the organization, agree that it is important enough to study and improve.
⎯⎯	⎯⎯	5. Enough managers, supervisors and operators in this area will cooperate to make this QFD Project a success.
⎯⎯	⎯⎯	6. This process is not currently being changed in any way, nor is it scheduled for change in the near future.
⎯⎯	⎯⎯	7. The QFD Project is clearly defined as one process that has well-defined starting and ending points.
⎯⎯	⎯⎯	8. This process is not being evaluated by any other group.
⎯⎯	⎯⎯	9. One cycle of the process is completed each day or two. Quick turnaround time is most important when selecting initial projects. Once a QFD team has some experience, longer, more complex projects can be attempted.
⎯⎯	⎯⎯	10. The QFD team's charter describes a problem to be studied or an improvement opportunity, not a solution to be tried.

2. Are all of the essential functions for this product/service represented on the QFD team, including customer service and field service
(if appropriate) and suppliers. This is especially important when either
or both of the following are true:

(a) The possibility exists of adding technical innovation to the prod-
uct/service, that is, Excitment Quality.

(b) The functions constitute a large proportion of the cost or value
to the ultimate customer.

3. Is the product or service in the mainstream of the company's
offerings? If not, will the company's culture afford a successful QFD
effort any "respect" if it is performed successfully on this product or
service? In other words, will personnel conclude from the choice of
the product or service to be redesigned using QFD that this is a trivial
methodology because it is used on a trivial product or service?

4. How much does your organization already know about the competitive aspects, both market and technical, of the product/service on
that you will be performing QFD analysis? Another reason to use a
product/service that is already in the market is that less effort will be
needed to collect the competitive data. The goal when *learning* QFD
is to not lose focus by getting involved in a lot of supporting activities
that, although they must be done, can consume time and resources
without providing a lot of understanding about the main QFD process.

5. Has this product/service been the object of any recent studies,
such as cost reduction or value engineering?

If so, are the results readily available to the QFD team? This will save
time and resources

6. If a redesigned product/service comes from the QFD team, will
the organization introduce it to the market in a timely manner? If not,
the message may be sent that QFD is a "nice" intellectual exercise
but not strategic or essential to staying competitive.

ACTION 8: ENABLE, EMPOWER AND ENCOURAGE

Enablement of a QFD team goes beyond understanding and agreement
with the project parameters (the GOOSE). It also includes provision
of the three A's:

- Awareness
- Appreciation
- Application

The three A's help a QFD Analysis Team to be aware of the expectations, appreciate what QFD can do to enhance a team's decision-making ability, and be able to apply QFD methodology to the selected project or study to be undertaken.

Empowerment includes the four P's:

- Power
- Permission
- Parameters
- Protection

The four P's help a QFD Analysis Team to be aware of the process of yielding power to act and decide by management permission to take whatever actions are necessary to achieve the project's parameters (the GOOSE) as well as to be protected from possible management intervention.

Encouragement of the QFD team refers to establishment and maintenance of an appropriate managerial climate, organizational culture and the spirit of cooperation between team members and non—team members.

ACTION 9: MANAGEMENT MAP PROCESS

To maximize QFD team output, management must provide both its commitment and involvement. An effective way of doing this is to use the MAP process:

Monitoring of the QFD team by management is necessary to ascertain team progress with respect to its anticipated completion date and resource consumption.

Auditing of the project parameters (the GOOSE) provides periodic updates on the team's progress using a bar chart analysis.

Probing demonstrates management's commitment to and involvement in various QFD projects.

Management should periodically probe the QFD teams with germane questions on

- Timeliness of the QFD process
  - Project parameters (the GOOSE)
  - Personnel appointments

- Assignment
- Research
- Customer identification process
- Timeliness of the Voice of the Customer acquisition process
- Demanded quality identification process
- Timeliness and accuracy of the competitive analysis process

Figure 19-6 is a template for the management guidance team to use in reviewing a QFD Project team's progress and the rationale for the project's selection.

The management guidance team needs to use the MAP process as a systematic approach to probe in-depth how the QFD team and the

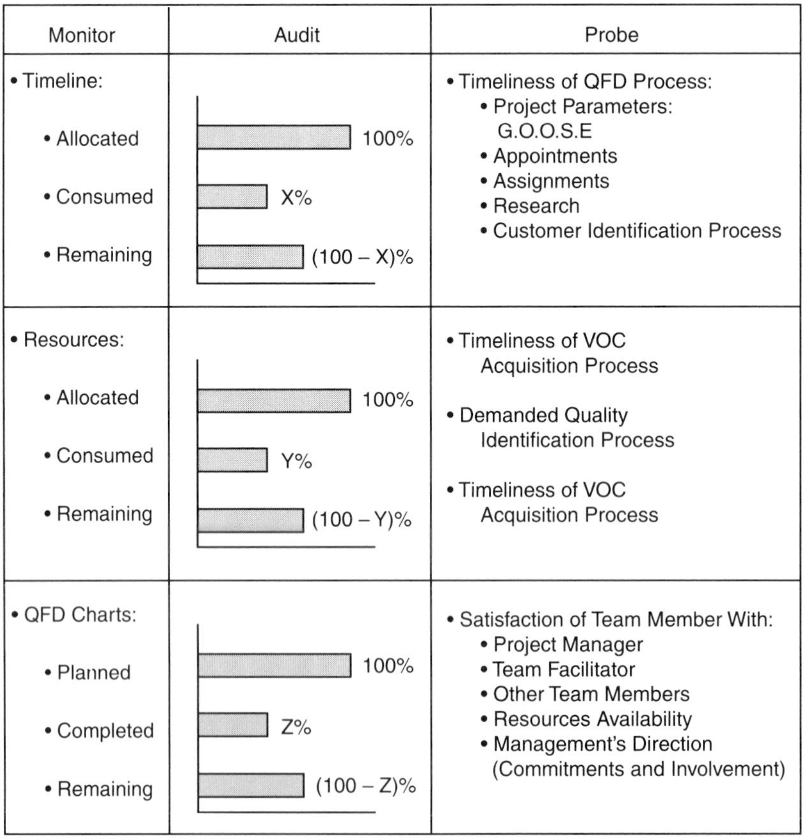

Monitor	Audit	Probe
• Timeline: • Allocated • Consumed • Remaining	100% X% (100 − X)%	• Timeliness of QFD Process: • Project Parameters: G.O.O.S.E • Appointments • Assignments • Research • Customer Identification Process
• Resources: • Allocated • Consumed • Remaining	100% Y% (100 − Y)%	• Timeliness of VOC Acquisition Process • Demanded Quality Identification Process • Timeliness of VOC Acquisition Process
• QFD Charts: • Planned • Completed • Remaining	100% Z% (100 − Z)%	• Satisfaction of Team Member With: • Project Manager • Team Facilitator • Other Team Members • Resources Availability • Management's Direction (Commitments and Involvement)

Note: X, Y, and Z will be approximately equal if the QFD Project is on schedule

Figure 19-6 Management MAP process project progress & status.

QFD process are really progressing. Besides asking the "right questions," the management guidance team also needs to

- Create and sustain the right cultural environment
- Maintain the right push by instilling a sense of urgency
- Provide the right training on a just-in-time basis
- Ensure the right resources are available as they are required

20

MANAGEMENT'S QFD REVIEW PROCESS

Every organization will want to generate its own, customized QFD Project Review Form for use by management. In an effort to help these organizations get started, we have developed the generic QFD Review Form shown in Figure 20-1. The QFD Review Form has a generic checklist of items and possible questions to be used as a guide. This review form should be customized to meet your organization's needs.

The form presented in Figure 20-1 is designed to help management focus on its roles and responsibilities in implementing and sustaining a QFD initiative. Management must define the product/service development process, set the broad strategic direction and monitor the external environment for possible shifts in the needs and requirements of their customers as well as the requirements of new or modified regulations.

Management must also ensure and provide the necessary resources so that access to real users is available to QFD Project teams. Access to real users and the information to be gleaned from them is essential because designers are poor surrogates for customers. Designers can be confused about the differences between the Voice of the Customer and the directions they receive from such sources as Request for Proposals (RFP) or Statements of Work (SOWs). Designers also can confuse elegant engineering with simple solutions that perform the desired functions for the customer.

Management must ensure that the QFD process stays focused on customer satisfaction at all times. QFD provides a means to keep the focus on the customer's needs, wants and desires to ensure consistent

QFD Project Title	Team Leader and Facilitator	Date of Review	Date of Last Review	Rationale for Project Selection
PF-3	J.Moran/J.ReVelle	5/7	3/10	Marketplace Leadership

Status	Improvements	Breakthroughs
● Team selected and in place ● Training completed ● Customer needs prioritized ● Planning Table completed ● Technical Characteristics I.D. ● Functions I.D. ● Charts completed (list out) ● Product Development Gates ● Rough FMEA	● Bottlenecks removed ● Speed Improvements ● Manufacturing flow ● Customer needs in design ● Handoffs improved ● Non-valued features removed ● Focused marketing data ● Packaging ● Instructions	● Better features ● Cost reductions ● Cycle time ● Simpler ● Improved reliability ● Surprise competition ● New features ● Enhancements ● New concepts

Questions	Support Needed	Status of Issues From Last Review
● What were the original project goals? ● Where are we today? ● Did you follow the QFD Process? ● Where did you deviate with the process? Why? Results? ● Did you receive adequate: · Training? · Facilitation? ● How did you collect the VOC? ● What criteria did you use to prioritize? ● What did the charts tell the team? ● What further information do you need? ● Where do you feel uncomfortable with the analysis? Why? ● What are the critical issues?	● What resources were requested? ● What resources have been used? ● What obstacles can we help with? ● What are your main roadblocks? ● What additional resources will you need? ● Any additional internal support needed?	● ● ● **Status of Issues From This Review** ● ● ●

Figure 20-1 QFD Project review form.

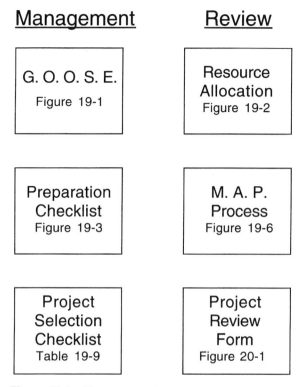

Figure 20-2   Management items versus review items.

customer satisfaction. Periodic reviews of QFD Projects by senior management have often been neglected and/or delegated. The review process is as important as the construction of the actual deployment chart(s). Review is the common thread between the initiation and the successful completion of a QFD Project. Review keeps the QFD process focused on the results and cycle time to completion.

The Management and Review Matrix shown in Figure 20-2 separates the items presented in Part IV into those that apply to management and those that are for project review.

21

GETTING TO KNOW THE CUSTOMER BEFORE DESIGN USING QFD

The *relentless pursuit of perfection* is how Toyota describes its approach to customer satisfaction: *relentless* in the sense that great is never good enough and the process of improvement will never end and *pursuit* in the sense that it will seek out what the customer wants, that which they do not even know to ask.

This relentless pursuit is not the same as eliminating dissatisfaction. Kano et al. (1984a) describe this in their article, "Attractive Quality and Must-be Quality." Figure 21-1 shows the Kano Model of Quality, with the three curves representing (a) Expected, (b) Normal and (c) Excitement Quality (see also Appendix D, Expanded Kano Model). The curve in the lower right quadrant (expected quality) suggests that no matter how extensively dissatisfiers are met, there is still not satisfaction. The center line (Normal Quality) indicates that there are known opportunities to create satisfaction by giving customers the things that they know they want. This is a powerful opportunity for improvement, but since these demands come from the customer, they are also easily known by competitors.

The top curve (Excitement Quality) in the Kano Model indicates those items that the customers are not aware of and so do not even know to ask for them. If a feature does not exist and the customer is not aware of the possibility for it, then there is no dissatisfaction. However, if such a feature does exist, there is potential for great satisfaction. QFD aims to uncover opportunities for Excitement Quality while making absolutely sure that Normal and Expected Quality issues are adequately addressed in the design process.

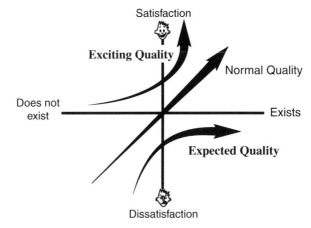

Figure 21-1   Kano Model of Quality.

   This goes beyond the traditional customer orientation that focuses on giving customers what they want or what the firm thinks they want. With the Voice of the Customer Table (VOCT) used at the front end of QFD, we can listen to and observe the customer using the product or service to learn what it does and then apply our capabilities to products and services based on their stated and unstated needs.

   This pursuit of customer satisfaction is more urgent when competitors catch up and are poised to surpass us. The design and development time available to deliver ever newer and more exciting products and services to the market is shrinking, making it difficult for companies to stay ahead. QFD gives an edge, but these days competitors are doing it too.

   The VOCT gives an added edge to QFD. It is a two-part tool to provide structure and a process for seeking out those items that the customer wants but also to surface other items that he or she may not even be aware of—to discover the customer's true needs.

VOCT PART 1

The first step is to gather data about how the product or service is being used or could be used by the customer (Nakui, 1991). These data come from interviews, questionnaires, observation of the customer's process in action (Nelson, 1992) or a relationships diagram of the customer's use (Zultner, 1992). At a minimum, look for who is using the product or service, what they are using it for and when, where, why and how they are using it. Record these data in the VOCT Part 1 (see Figure 21-2). New uses may also be suggested by customers or by the QFD team. For firms that supply components to be used by Original Equipment Manufacturers (OEM), VOCT data often take the form of specifications or an industry standard. In these cases it is especially important to understand how both the OEM customer and the end user will use the product, since customers could still be dissatisfied even when specifications and requirements have been met (Mazur, 1991).

VOCT PART 2

In Part 2, the Voice of the Customer is reworded into demands taking into account all of the uses described in Part 1. Demands take on many forms such as demands for quality, performance, low price, long life, safety or low environmental impact. The QFD design team must interpret these various demands from the Voice of the Customer, the usage and the operating environment. Be creative in interpreting and do not worry about far-fetched demands; the customer can be asked to prioritize these using the importance ranking in the quality planning section of the House of Quality; preposterous ones will usually rank low. You might discover something new should one such demand be ranked high!

After the demands have been extrapolated from the Voice of the Customer, a sort is done to make later prioritization more effective. This is needed because the demands are usually a mixture of demanded quality, quality characteristics, functions, methods, reliability and other issues, and the prioritizing must take place between items similar in

Voice of the Customer	Use					
	Who	What	When	Where	Why	How
Easy to find during night time power failure.	Adults; kids	See during power failure	Night	House; basement	See in dark; check fuses	Hold in hand; set on surface

Figure 21-2   VOCT Part 1 for a flashlight.

nature. Specific matrices designed to relate each of these will be shown later. The types of demands are defined below.

Demanded quality is a qualitative expression of the benefit the product gives the customer. These should be concise, singular and positive. Example: *can hold easily.*

Quality characteristics are quantifiable, measurable or controllable expressions for achieving demanded quality. Example: *diameter.*

Function is what the product or service does or its purpose. Based on value engineering, functions are expressed as a verb plus an object. Example: *maintain aiming.*

Reliability is the expected life of the product. Often included here are the product failure modes or complaints. Example: *doesn't work.*

Other items can include price, safety, environmental friendliness, and so on.

In Figure 21-3 the reworded data items are entered into the appropriate columns. Those that are already expressions of demanded quality are placed in the demanded quality column. Quality characteristics, functions or methods are clues to hidden, true needs. We should first consider why the customer might want something like this. That is, what benefit does the customer derive? These should be expressed as demanded qualities and placed in the demanded quality column (see Figure 21-4). Reliability, failure modes and other items should not be converted back to demanded qualities since they are so critical they would likely overwhelm the demanded quality items during prioritiza-

Reworded Demands	Demanded Quality	Quality Characteristics	Function	Reliability	Other
Can hold easily	Can hold easily				
Can use hands free	Can use hands free				
Maintain aiming			Maintain aiming		
Fits in drawer		Diameter			
Always ready to use				Does not work	

Figure 21-3   VOCT Part 2 for a flashlight.

Reworded Demands	Demanded Quality	Quality Characteristics	Function	Reliability	Other
Can hold easily	Can hold easily				
Can use hands free	Can use hands free				
Maintain aiming	Can see easily		Maintain aiming		
Fits in drawer	Can store easily	Size			
Always ready to use				Does not work	

Figure 21-4   Analyzing and organizing demands.

tion. Specific matrices for these will be used after the House of Quality Matrix.

When this analysis is complete, the demanded quality column will then be grouped in an affinity diagram for further deployment. Quality characteristics, functions and methods can be included in appropriate matrices, as can reliability and other terms. Figure 21-5 is a table of matrices that use the VOCT Part 2 items. Nomenclature is derived from the Matrix of Matrices developed by Bob King (1989a) of GOAL/QPC.

	DQ	QC	F	R	M
DQ		A-1	B-1	D-1	
QC		A-3	A-2	D-3	* C-1'
F				D-2	C-1"
R					D-2'
M					

Figure 21-5   Table of matrices for VOCT Part 2 elements. The C-1', C-1" and D-2' matrices come from an extension of the Matrix of Matrices to specify matrices that transform quality characteristics, function or reliability, respectively, into methods.

VOCT BENEFITS

- True customer needs are identified by translating verbalized and observed data.
- New demands can be developed based on analysis of current and potential uses.
- New markets and new products can be developed.
- Resulting matrices are smaller and better organized.
- Prioritizations resulting from matrices are not skewed by inappropriate entries.

QFD Projects that have used VOCT have found good results. While the process adds an additional step, it provides a structure for team members to work together on analyzing customer data and makes resulting matrices smaller, better organized and easier to work with. Some teams have said that the VOCT exercise was the most valuable part of QFD.

22

PLANNING FOR PRODUCT, SERVICE OR PROCESS DEVELOPMENT USING QFD

Another management responsibility with respect to the application of QFD exists in the area of planning the strategies and concepts for new products, new services and new processes.

The development of a continuous and steady stream of innovations is the life blood of an organization. Organizations remain viable in the marketplace by introducing timely and value-adding products, services and processes in addition to their current ones. This helps them retain their current customers and attract future ones. QFD concepts and tools can be utilized during the development phase to ensure a smooth transition to the team that will take the concept from an approved idea to a deliverable product or service.

Turning enhanced technologies into marketplace winners is a management responsibility of the highest priority. The challenge to management is to develop a consistent approach to monitor, review and direct the technology of the organization from the idea phase to commercial development to marketplace delivery. This approach has to be clear in its intent since it will become a guiding policy to produce valued and differentiated product offerings for the organization's customer groups. A consistent planning, monitoring and reviewing approach is a necessity for organizations seeking to establish quality systems based on ISO 9000 and beyond.

The process described in this chapter uses QFD to enhance a strategic technology development and exploitation process. A technology development and adoption strategy must have clear screening criteria, integration of market research to capture expressed and latent customer needs, a cost–benefit analysis with respect to the organization and its customers and predetermined evaluation points to ensure that the proposed commercialization of a technology is feasible and cost effective. This process will help to couple the core technology foundations of the organization with the market needs. In addition, it will produce strategic technology investment opportunities that lead to growth and profitability, both in the short and long run.

QFD in the new-concept stage is quite different than in the continuous-improvement or next-generation product/service stage. Figure 22-1 contrasts this relationship of QFD with new concepts and next-generation products/services.

In Figure 22-1, the top half depicts the present, near-term activity in support of achieving customer satisfaction. The bottom half deals with the needs of the future. These often latent customer needs are difficult to collect and analyze but well worth the investment since they hold the keys to new products and services. Developing new concepts requires an inventor's mentality with a long-term time frame rather than a designer's or engineer's mind set, which is short term and task oriented.

The development of a steady stream of innovations requires the organization to partner with the customer to build a commitment to one another rather than just focusing on satisfying the customer's current needs. Customer commitment is different than customer satisfac-

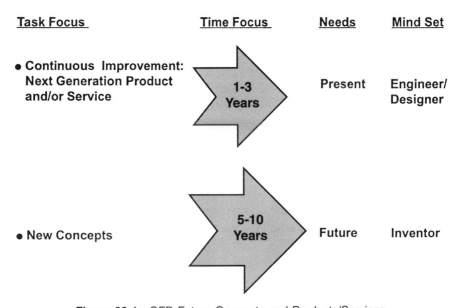

Figure 22-1 QFD Future Concepts and Products/Services.

tion since with commitment comes a sense of loyalty and interdependence. This commitment is missing with customer satisfaction. Customers can be satisfied, but not loyal or dependent, since they have many options with which to satisfy their needs.

The use of QFD in the deployment phase is shown in Figure 22-2. The main QFD tools and matrices used for this purpose are the Voice of the Customer Table (VOCT), Chart E-1, Chart C-1 and Chart A-1 (Planning Table) (see Appendix A, QFD/Pathway and Matrix Interrelationships). To use QFD in the development phase, there must

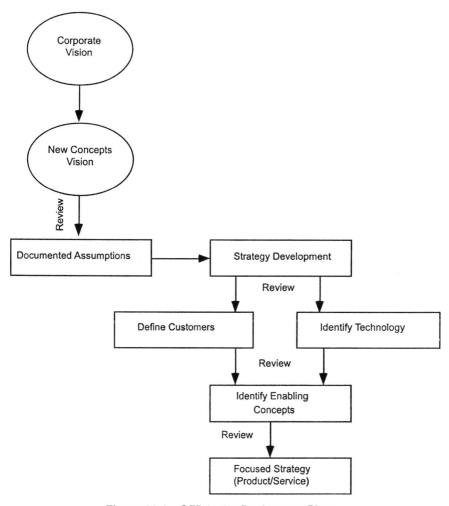

Figure 22-2   QFD in the Deployment Phase.

be an overall corporate business strategy that is clearly defined. The development team must translate the corporate business strategy into a business unit mission that projects out 5–10 years. The mission is supported by documented assumptions about the economy, the industry, the competition, the future needs of customers, the market size, the likely regulatory environment, and so on. Once these assumptions are documented, the development team produces a series of strategies to follow to achieve marketplace leadership.

The process shown in Figure 22-2 includes review cycles. Too often new concept teams are guilty of springing new ideas on the corporate approval group. As a result of this sudden, unplanned approach, the decision regarding approval to go ahead and pursue a given idea is often delayed until the corporate approval group has had a chance to educate itself in the hows and whys of the idea. By having a planned and structured corporate review process for new ideas and concepts, delays can be minimized. Decreasing the potential for delays in approval increases the chance that the organization can be first to market with the advantage of greater volumes of product sold earlier and over a longer period of time with the premium pricing usually associated with first-to-market products/services. The new concept review cycle should logically lead into and be an integral part of the development review and funding process cycle. Market research and strategy testing should be an ongoing process, not just an event associated with one specific new concept development initiative. The resulting, constantly up-dated database provides more information to the corporate approval group, especially with respect to trends and areas that are just beginning to grow out of mature products/services. And due to the greater information available, the corporate approval group has much greater confidence that the idea or concept being reviewed is based on sound research and valid assumptions.

Each strategy produced should be checked against the documented assumptions to ensure that it is aligned with the basic tenets of the assumptions. This alignment can be accomplished by utilizing the Alignment Matrix shown in Figure 22-3. The traditional QFD relationship matrix symbols and associated point weights can be applied to facilitate the analysis. A Pareto diagram of the strategy scores indicates which are top ranked and require development.

The strategies chosen from the Alignment Matrix can now be compared to future customer needs. Future customer needs are sometimes referred to as latent needs since they may not be apparent to current customers or even to potential future customers. There are many different avenues available to gather these latent needs. Some methods to employ are

ASSUMPTIONS

	A-1	A-2	A-3	A-4	A-5	A-6	Total	%	Rank
S-1	◎	△	◯	◎		△	23	X%	1
S-2									
S-3									
S-4	△	△	◯		◎	△	15	Y%	2
S-5									
Total							Σ	100%	

STRATEGIES

Figure 22-3 Alignment Matrix.

- Market trend analysis
- Patent searches
- Working with customers at their location
- Surveys
- Blind focus groups
- Industry experts
- Thought leaders
- University or national laboratory researchers
- Independent inventors
- Futurists
- Benchmarks (process/product focused)
- Strategic (industry-focused) benchmarks

The future customer needs can be compared against the selected product strategies to determine which strategies are worthwhile to pursue and perhaps develop into a product or service. Figure 22-4 presents an evaluation matrix to assist in this evaluation. The same relationship matrix symbols are used as in Figure 21-3.

FUTURE CUSTOMER NEEDS

	N-1	N-2	N-3	N-4	N-5	N-6	Total	%	Rank
S-1	◎	○	○	◎	◎	◎	42	X%	1
S-2									
S-3									
S-4									
S-5	△		○			◎	13	Y%	2
Total							Σ	100%	

STRATEGIES

Figure 22-4   Evaluation Matrix.

Once the future customer needs are understood and the appropriate product/service strategies are selected, the development team uses the C-1 and E-1 QFD matrices (see Appendix A, QFD/Pathway and Matrix Interrelationships) to assist in the selection of concepts and technologies to enable the product/service to be developed.

A new concept must be developed into a plan that documents all the necessary information needed to secure approval to develop a new concept into a tangible product or service. The development planning document is a must for any organization's technology planning process. It provides the mechanism to document the initial and updated assumptions used to prove that the commercialization of a technology merits investment of the organization's limited resources.

To be competitive in today's global marketplace, organizations must prioritize and effectively manage their R&D strategies. To accomplish this, an organization needs to focus on a few key core competencies that can create a strong competitive advantage. This focus on core competencies leads to the development of many new and complimentary products and services that provide growth and increased profitability to the organization.

An organization needs a development plan that integrates its core technology and strategic focus or intent with the market's needs. This framework or development plan for new products and services helps management reduce the risk to the organization associated with R&D investments. The risk is reduced by a careful and consistent evaluation during the development planning process to ensure that each proposed project is evaluated by the same standards and is truly focused on the marketplace and the customers and is aligned with the organization's strategic focus or intent. This alignment encourages R&D to focus and prioritize development efforts, which results in a balance between short-term and long-term development programs.

To assist in achieving these goals, a generic, new product/service development plan is detailed below. This plan is focused on the following major topic areas:

- Concept overview
- Market analysis
- Competitive analysis
- Organizational impact
- Customer analysis
- Financial justification
- Recommendations and conditions

From the authors' experience, these points of focus are the most common areas of investigation. They should be modified to fit an individual organization's needs.

A consistent and disciplined approach to developing, selecting and monitoring/reviewing the development of new products and services is a plus for organizations moving toward or adopting an ISO 9000 oriented quality management system.

Generic Strategy and Feasibility Product and Service Development Plan

Note: Blank forms for each section are located in Appendix C.
Section 1: Concept Overview
        A. Concept description:
           A clear, concise description of the concept being advanced for consideration. Identify the intended customer group(s). Indicate if this concept will be a market leader or follower. Indicate the differentiation of this concept

from existing products and/or services (which functions/ benefits are new or different).

B. Core assumptions:

Why pursue this concept? List the key assumptions and trends plus the supporting documentation, data and/or references that support the concept being advanced for consideration.

C. Issues to be researched:

At the initial stage of the concept advancement, what areas should be further researched to validate the need and gain approval of this concept for a company product or service?

Issues	Research Methodology	Findings
•		
•		
•		
•		
•		
•		

Section 2: Market Analysis

A. Best estimate of the date of market entry:

Quarter:_____   Year:_____

1. Issues that would delay the entry date and their impact on that entry date.

Issues	Length of Delay If Unresolved
•	
•	
•	
•	
•	
•	

2. Is there a market window? If yes, describe it.

Document the timing and duration of a potential market window that must be met. Document the resulting consequences if this market window is missed.

B. Market scope:

1. Documentation for estimates:

List sources and references for estimates to be made concerning the market for this concept.

2. Market size:

	Domestic		International	
Now	$$ _____	and Units _____	$$ _____	and Units _____
5 Years	$$ _____	and Units _____	$$ _____	and Units _____
10 Years	$$ _____	and Units _____	$$ _____	and Units _____

3. Market target location list with dollar and volume projections:

Target Location	Domestic/ International	Dollar Potential	Volume Potential
•			
•			
•			
•			
•			

4. What approach will be used to attain this market share?
   Describe the activities that will be undertaken, such as promotions, advertising (what and where), trade exhibits, pricing, appointing agents, and so on. Estimate the market share impact of each activity.

C. Pricing strategy:
   Describe the pricing strategy to be followed initially and as the market matures. Describe any discounting strategies. Describe any pricing leadership strategy that may prevent the competition from pursuing this concept.

Section 3: Competitive Analysis
   A. Competitive position in target markets in dollars and units

Competition	Now ($/Units)	5 Years ($/Units)	10 Years ($/Units)
•			
•			
•			
•			
•			

B. Anticipated competitive reaction:
Once the competition knows we are pursuing this concept, what might their reaction be?

Competitor	Reaction
•	
•	
•	
•	
•	

C. Is the competition working on this concept now? If yes, what is their anticipated market entry date?

Competitor	Entry Date
•	
•	
•	
•	
•	

Section 4: Organizational Impact
    A. Value and risk analysis:
        1. Describe the strategic value to the corporation:
        Is this concept a next-generation product/service or a breakthrough in technology? How does this concept help in portfolio balancing?
        2. Risks to the corporation concerning pursuit and non-pursuit of this concept:
        Indicate potential positive and negative trends and their probability of occurrence.

Trend	Positive/ Negative	Probability of Occurrence
•		
•		
•		
•		
•		

3. Overall probability of success of this concept in the marketplace and the assumptions to support it.

B. Organizational impact:

Describe new processes, skills, resources, facilities, and so on, required to fully develop this concept into a product/service. When will each be required and what are the costs?

Required	When	Cost
•		
•		
•		
•		
•		
•		

C. Regulatory/environmental requirements and impacts:

Describe any regulatory or environmental impacts to the corporation if this concept is pursued.

D. Patent position:

1. Results and data of current patent search
2. Potential of exclusivity

Section 5: Customer Analysis

A. Customer perceived value:

Why would the customer purchase this concept? What previously unsatisfied want or need does this concept satisfy? What would the customer be willing to pay for this concept?

B. Customer requirements:

List the customer requirements obtained from the marketing research studies conducted. Break down the requirements into (1) Basic Quality (dissatisfiers/musts), (2) Performance Quality (satisfiers/wants) and (3) Exciting Quality (delighters/wows) categories. Indicate the satisfaction target to be achieved for each customer requirement. Detail the expected weight, throughput, setup time, reliability, operating costs, features/benefits, and so on. In addition, indicate how this satisfaction target will be measured and monitored.

Customer Requirement	Satisfaction Target	Measurement Strategy
•		
•		
•		
•		
•		

C. Internal requirements:
   List any internal organizational requirements that may be important for smooth product functioning or service delivery such as safety, environmental, cost targets, timing, and so on. Indicate the satisfaction target to be achieved for each internal requirement and indicate how this satisfaction target will be measured and monitored.

Internal Requirement	Satisfaction Target	Measurement Strategy
•		
•		
•		
•		
•		

Section 6: Financial Justification
   A. Financial analysis:
      Include results of usual financial analyses: Breakeven Point (BEP), Return on Net Assets (RONA), Return on Investment (ROI), Net Present Value (NPV), Payback Period, Gross Margins, Capital Investment, Development Costs, Internal Rate of Return (IRR), and so on.
   B. Preliminary feasibility and development budget:
      The project leader should detail a preliminary feasibility and development budget with the necessary assumptions that alert the management team to the size and chronology of the proposed investment of the organization's resources.

Section 7: Recommendations and Conditions
   A. Recommendations:
      What does the New Concept Team recommend to the Approval Team?

B. Approval/conditions:
   The Approval Team details the approval and any conditions that they may wish to attach at this point.
C. Review dates:
   Develop the future schedule of review dates for updates and modifications to this document and the scope of the project.

23

ALTERNATIVE
APPROACHES TO QFD

Knowledgeable QFD practitioners are not inflexible when applying QFD to their projects. Rather, they can use either of two well-known approaches to QFD—the 4-matrix model or the 30-matrix model described by the Matrix of Matrices—as well as variations on these approaches.

The 30-matrix approach (see Figure 23-1) is most successful for projects that require more detailed understanding as a result of using QFD. The 4-matrix approach (see Figure 23-2) is a "Reader's Digest" version of the 30-matrix approach. When you consider all the multilevel analyses and details associated with deploying the Voice of the Customer, it becomes apparent that for many projects four matrices are just not enough.

For example, the four-matrix model provides only one matrix for production planning while the Matrix of Matrices contains six matrices just for manufacturing. In other areas as well, the Matrix of Matrices provides more depth than the four-matrix model. And yet we recommend that QFD neophytes as well as seasoned QFD practitioners working on reasonably straightforward assignments make use of the four-matrix model.

As a matter of interest to our readers, ReVelle and Moran teamed together to make a QFD-focused presentation at a large (about 1000 attendees) international conference in 1989. As part of the presentation, they did a real-time survey of their audience to ascertain the proportion of attendees who used the four-matrix and Matrix of Matrices approaches to QFD.

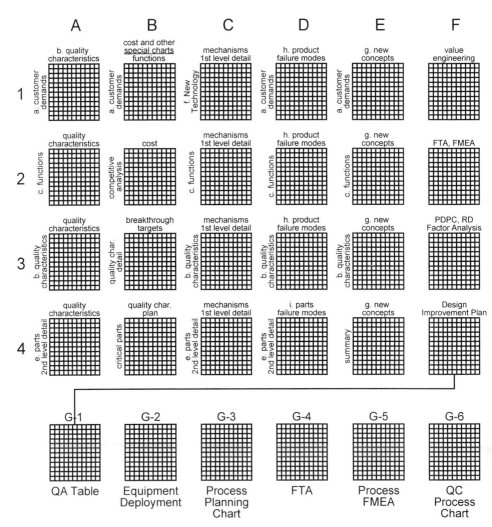

Figure 23-1 The Thirty-Matrix Approach.

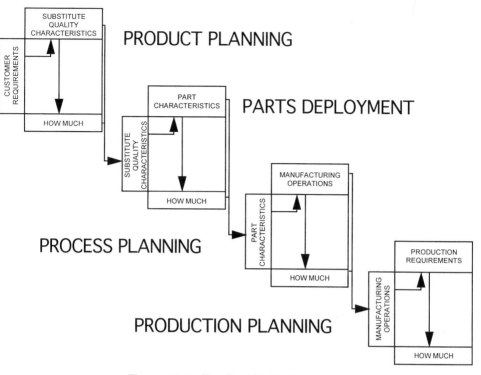

Figure 23-2   The Four-Matrix Approach.

They were surprised to learn (although in retrospect it should not have been a surprise at all) that while 10% of the QFD practitioners in the audience used the four-matrix model exclusively, another 10% used the Matrix of Matrices approach exclusively. The surprise was that the remaining 80% used an integrated approach combining the best features of both models. One such integrated model is detailed in the book *Facilitating and Training in Quality Function Deployment* (Marsh et al., 1991). This integrated model is built around Deming's Plan, Do, Check, and Act (PDCA) cycle. It starts at the marketing level with the Voice of the Customer Table and works its way through to element design using 13 of the QFD matrices from the Matrix of Matrices.

Since that time, it has been confirmed that an overwhelming proportion of QFD practitioners select the matrices and software that best fit their needs and run with whatever ball they choose to carry. Every player has his or her own set of rules that are established to fit their organizational culture, their management structure/bureaucracy, their projects, their objectives, their budget and whatever other factors may influence their decisions.

Fortunately, QFD is a robust tool, that is, it is difficult to do it wrong. However, sufficient training on both approaches, facilitation experience and supportive, involved management improve the results dramatically.

Sometimes it is simply not possible to do a comprehensive QFD project involving any combination of the Matrix of Matrices and the four-matrix approaches. For these cases, Richard Zultner, a student of Akao, has developed a streamlined approach to QFD, called "Blitz QFD," that is upwardly compatible to comprehensive QFD. Blitz QFD was developed for QFD teams that have constraints on time, people and money. Rather than give up on QFD totally, Blitz demonstrates how to select and deploy only the top most important ranked customer needs.

Beyond the requisite planning of your QFD process, there are seven steps to applying Blitz QFD:

1. Gather the voice(s) of your customer(s).
2. Sort the verbatims you receive from the customers.
3. Structure the customer needs.
4. Analyze the customer needs structure.
5. Prioritize customer needs.
6. Deploy the prioritized customer needs.
7. Analyze only the important relationships in detail.

These are the seven steps to a streamlined start with QFD. It is useful to note that although comprehensive QFD makes use of matrices, Blitz QFD does not. Matrices are excellent when you want to examine in detail a large number of interactions between two dimensions. If you have the time, resources and money, then do comprehensive QFD. But if not, then there are more efficient ways to get at and handle that small number of highest value items (which is all we have time to do on a constrained project). Note that just doing a matrix does not mean you are doing QFD! And not using matrices in Blitz QFD does not mean you are not doing QFD.

The approach taken by Blitz QFD is actually a return to the very roots of QFD. [For more details on how and why QFD was initially developed see Mizuno and Akao's (1994) book, *QFD: The Customer-Driven Approach to Quality Planning and Deployment.*]

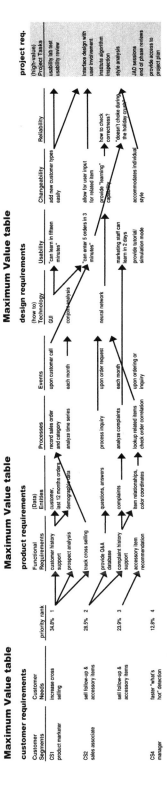

Figure 23-3 Customer Voice Table.

In the first step of Blitz QFD, we consider the various categories of customers and determine which types of customers are most important. For Blitz, we only focus on the most important customers. We ask ourselves,"Who must we satisfy if we are to succeed?" We should also ask, "Who must not be dissatisfied or we fail?" Once we have identified the customer group(s) most important for our success, we conduct our information gathering at the customer's site, where the product (or service) is actually used. It is important that we go to where "the action is," observe the product/service in actual usage and note the environment and actual circumstances prior to, during and after usage. It is essential to understand what the customer does and why as well as what the customer does not do and why not. Discuss with the customers what their problems are, solicit their comments and observations and record them exactly as the customer states them (verbatim).

affinity diagram
customer needs

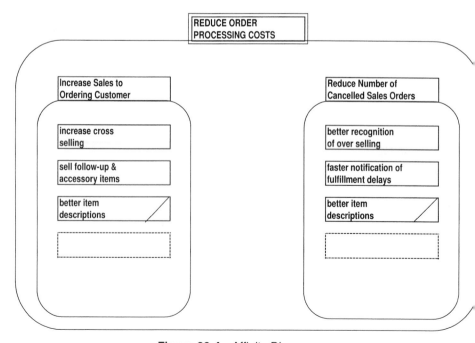

Figure 23-4   Affinity Diagram.

The second step is to sort the verbatims that we have received from the customers. The Customer Voice Table (CVT) (see Figure 23-3) assists in the process of organizing the verbatims, allowing you to see what comments you do have and to note those areas where comments are missing. In the CVT, those concerns that are most important to customers are the columns of the table, and these are the dimensions critical to the success of our project. Elements of the CVT vary with the situation but often include many of the following:

Customer: characteristics, problems and needs

Product: characteristic and capabilities, functions (for hardware), tasks (service), objects (for software)

Design: reliability, technology, cost (price), parts

Other: project, social, organizational

The third step is to organize and structure the customer needs, which is done using an affinity diagram (see Figure 23-4). The team arranges

hierarchy diagram
customer needs

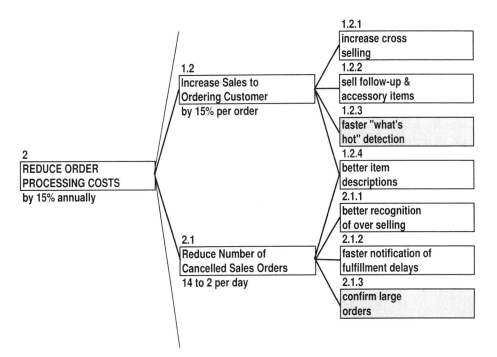

Figure 23-5   The Hierarchy Structure.

the customer needs into their "natural" groupings. Once the various customer needs have been clustered, the headings are analyzed.

The fourth step is analyze the customer needs structure with the use of a hierarchy structure (see Figure 23-5). The emphasis in this step is on understanding the customer's thinking, finding his or her unstated requirements, quantifying the customer's needs and then prioritizing them (next step).

The fifth step is prioritizing the customer's needs. The Analytic Hierarchy Process (AHP), a pairwise comparison approach, allows the team to accurately prioritize the needs using ratio-scale numbers (see Figure 23-6).

Once the needs are prioritized, then in the sixth step, a plan is developed to deploy the most important identified customer needs. By analyzing the CVT, you can determine which are the most important contributors to the most important customer needs. In some instances, there may be cases where the contributors are not listed, and they will need to be added. From this exercise, a maximum-value table (see Figure 23-7) is created. Using the CVT, those items that contribute

pairwise evaluation matrix

analytic hierarchy process

Row Average of Normalized Columns (RANC) method

customer needs items	1 increa	2 sell f	3 faster	4 better	1	2	3	4	row total	row avg.
					normalized columns					
1 increase cross selling	1	2	3	9	0.51	0.58	0.42	0.36	1.87	0.468
2 sell follow-up & accessory items	1/2	1	3	8	0.26	0.29	0.42	0.32	1.29	0.322
3 faster "what's hot" detection	1/3	1/3	1	7	0.17	0.10	0.14	0.28	0.69	0.172
4 better item descriptions	1/9	1/8	1/7	1	0.06	0.04	0.02	0.04	0.15	0.038
total	1.9	3.5	7.1	25.0	1.0	1.0	1.0	1.0	4.0	1.0

extreme	9.0	1/9	0.111	relative
	8.0	1/8	0.125	judgment
very strong	7.0	1/7	0.143	scale
	6.0	1/6	0.167	
strong	5.0	1/5	0.200	
	4.0	1/4	0.250	
moderate	3.0	1/3	0.333	
	2.0	1/2	0.500	
equal	1.0	1/1	1.000	

Figure 23-6   The Analytic Hierarchy Process.

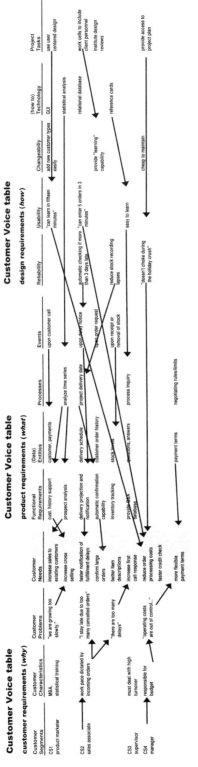

Figure 23-7 Maximum value table. On the CVT, those items that contribute most to satisfying the most important customer needs, are the maximum value items.

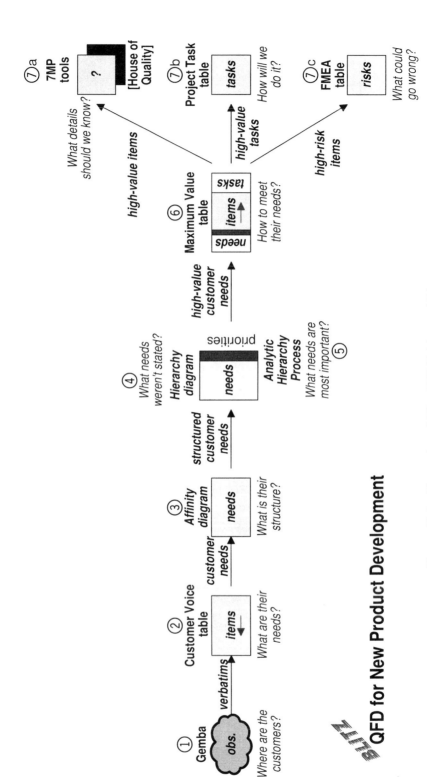

Figure 23-8 The series of Blitz QFD Tables and Diagrams.

the most to satisfying the most important customer needs are the maximum-value items.

In the last step, analyze only the important relationships in detail—and only to the extent that is warranted. Stay focused on the high value items and only explore to the depth necessary. Look at the details in one column, or the interactions between two columns.

The Blitz QFD series of tables and diagrams is shown in Figure 23-8. It shows a sequence of enquiry, analysis and conclusions which are common to comprehensive QFD, but due to the constrained resources and limited subject matter (only the most important customers and only their most important needs), you don't need to use matrices. However, the usefulness of the derived information is still there and can serve as the foundation to a comprehensive QFD project.

V

APPENDIXES

APPENDIX A

QFD/PATHWAY AND MATRIX INTERRELATIONSHIPS

THE REASON FOR QFD/PATHWAY

Determining the matrices that will be necessary to work through for a particular QFD Project depends on the project's goals (desired outputs) and what intermediate matrices will be necessary to accomplish the goals. The simplest sequence of matrices, of course, is T. Fukuhara's four-matrix series popularized by the American Supplier Institute (see Figure A1). That series may or may not serve the purposes of the particular QFD Project being considered. It is often the series that is taught when teaching the QFD methodology because it fits many circumstances and is intuitive.

An approach that offers many more choices and options is Y. Akao's Matrix of Matrices (see Figure A2) popularized by GOAL/QPC and Bob King (1987c, 1989a) in his book, *Better Designs in Half the Time*. The difficulty with using the Matrix of Matrices is that to the uninitiated it can be quite daunting, and it may become more of a maze of mazes than a Matrix of Matrices. To the rescue comes a software package, QFD/Pathway, which helps guide the QFD team through the maze to reach its goals in an effective manner. This software package is included as an add-on to this text.

QFD/Pathway is a totally new, user friendly, expert system designed by experienced QFD practitioners. It provides needed assistance in moving a QFD Project from the House of Quality (HOQ, or A1 matrix) to the next steps. (It also addresses whether the team wants to use the Voice of the Customer Table to capture and refine customer inputs before creating the House of Quality.) The next steps may include one or more other matrices that are graphically identified in the Matrix of Matrices (Figure A2), as well as other more recently developed matrices

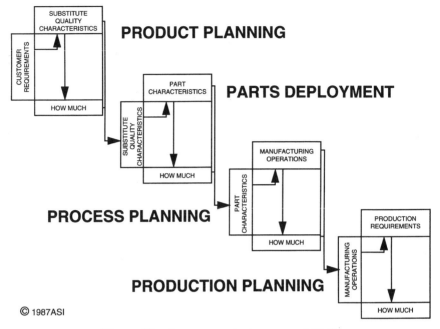

Figure A1   Four conceptual phases of QFD.

not included in the Matrix of Matrices. By some counts there are nearly 40 matrices from which to choose as a team prepares to move its project beyond the House of Quality (A1 matrix).

QFD/Pathway assists an organization to move from translation of its project goals and strategies to selection and sequential deployment of QFD matrices. Just as the American Automobile Association (AAA) provides its members with graphic assistance in the form of Trip Tik™ strip maps (see Figure A3) to show the fastest, cheapest or most picturesque path to follow from one geographic location to another, so does QFD/Pathway provide step-by-step assistance in the form of easy-to-follow documentation that leads a team from matrix to matrix, ultimately concluding its trip with deployment of the optimal QFD matrix set (see Figure A4). AAA strip maps are limited to between-city information, leaving within-city details to larger street maps. So too does QFD/Pathway focus strictly on the sequential deployment of QFD matrices, leaving the analysis of each matrix, to the discretion of the user.

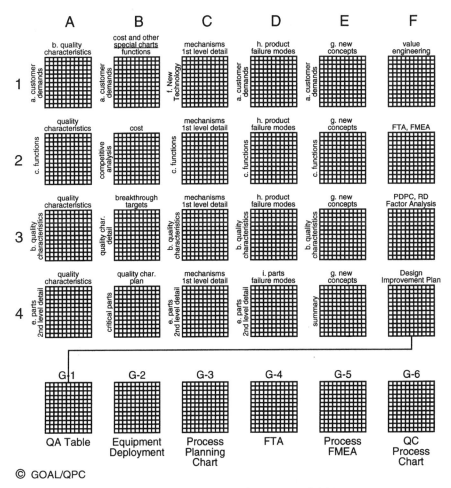

Figure A2 The Matrix of Matrices of QFD.

The selection of the optimal QFD matrix set is dependent on the situation, that is, whether the team is analyzing a project involving hardware, software or service. For example, while one QFD project may involve hardware and require selection and sequential deployment of four specific matrices, another hardware project may require use of 10 matrices, some which are common to the first project and some of which are not. Further, because of situational differences between hardware, software and service, some specific matrices vary in content and format depending on the focus of the team. QFD/Pathway offers

Figure A3   Trip Tik™ map.

its users the option of working with one of three alternative paths: hardware, software and service.

The typical user of QFD/Pathway is most likely to be a team of QFD practitioners with a blend of QFD experience levels varying from little to none all the way to fully experienced facilitators and team leaders. Typical users may be documenting their efforts manually, with generic spreadsheets or with an existing QFD software package such as *QFD/Capture* (from ITI; see Appendix B) or *QFD/Designer* (from ASI; see Appendix B). Whatever form of documentation is employed,

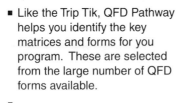

- Like the Trip Tik, QFD Pathway helps you identify the key matrices and forms for you program. These are selected from the large number of QFD forms available.

- These key matrices and forms will ink the elements you specify

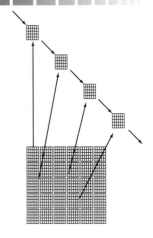

Figure A4   QFD/Pathway.

the typical team is stymied in its efforts due to uncertainty or ignorance regarding where to go from the House of Quality. This blockage is eliminated through the consistent use of QFD/Pathway from the outset and is repeated, as needed, throughout the life of the project.

In addition to assisting the QFD team as it pursues the relationships between the various matrices to reach its goals, QFD/Pathway can be called upon to print the sequence of matrices used and their interrelationships. This serves both as a forecast of budget and schedule requirements as well as a record of the QFD team's efforts. It provides an excellent basis for communicating the "journey" the QFD team has taken to get its results and provides the background for high-level briefings on the QFD Project, the scope it covers and where the answers the QFD effort provided came from. If other persons get interested in taking the same journey, they will have a better understanding of what to expect, in terms of both the needed inputs and the results and how to attain them. This will form the baseline upon which they may be able to improve.

SYSTEM REQUIREMENTS AND INSTALLATION FOR QFD/PATHWAY

The QFD/Pathway software design and production was authored by Jack B. ReVelle and R. Alan Kemerling. It is available for IBM PCs and compatibles and requires a 386 or higher CPU and approximately 1 megabyte of space on the hard drive. It works best with 16 or more megabytes of RAM. To install the QFD/Pathway software on your hard drive, insert the QFD/Pathway floppy disk in drive a (or b depending on how your computer is configured), go to Program Manager in Windows, click on File, go to Run and type in a:setup (b:setup if floppy disk was placed in drive b) and hit Return.

APPENDIX B

QFD INFORMATION SOURCES

ORGANIZATIONS THAT OFFER PRODUCTS OR SERVICES RELATED TO QFD

American Society for Quality
611 E. Wisconsin Avenue
P.O. Box 3005
Milwaukee, WI 53201-3005
Phone: 800-248-1946, also 414-8575; Fax: 414-272-1734
email: asqc@asqc.org
Website: http://www.asqc.org
Contact: Programs and Education Manager

The ASQ is an organization dedicated to the advancement and deployment of knowledge about the science of quality control. The society publishes four journals that contain QFD-related articles, offers an ASQ Net and Bulletin Board and technical committees and conducts public courses on QFD.

American Society for Quality (ASQ) Quality Press
P.O. Box 3066
Milwaukee, WI 53201-3066
Phone: 800-248-1946, also 414-272-8575; Fax: 414-272-1734
Website: http://www.asqc.org
Contact: John R. Hackl, Publisher

Quality Press provides a comprehensive catalog of resources on the entire field of quality. The material in the catalog are abstracts of practical, peer-reviewed information tailored to particular needs and levels. They offer a number of QFD and related books for the beginner as well as the advanced practitioner.

American Supplier Institute (ASI)
17333 Federal Drive, Suite 220
Allen Park, MI 48101
Phone: 800-462-4500, also 313-336-8877; Fax: 313-336-3187
email: pat@amsup.com
Website: http://www.amsup.com
Contact: Dave Verduyn, V. P.

The ASI pioneered the implementation of QFD and Taguchi methods in the United States. ASI provides on-site consultation and training in QFD. ASI has developed three different approaches for QFD focusing on product design, services and processes and continuous batch processing. ASI offers books and software to assist an organization with its implementation of QFD.

GOAL/QPC
13 Branch Street
Methuen, MA 01844-9922
Phone: 800-643-4316, also 508-685-6370; Fax: 508-685-6151
email: service@goal.com
Website: http://www.goalqpc.com
Contact: Bob Page, Manager, Learning Materials

GOAL/QPC, a nonprofit TQM organization and publisher of the Memory Jogger series, offers books and on-site training in QFD.

IBM Consulting Group
11400 Burnet Road, M.S. 4502
Austin, TX 78758-3493
Phone: 512-838-7025; Fax: 512-838-8400
email: cadiano@vnet.ibm.com
Contact: Cindy Adiano, Quality Consultant

IBM Consulting offers consulting and private workshops on Design QFD and Dynamic QFD as well as two software packages: IBM Strategic Pointer 2000/2™ with customer, supplier and statistical process control modules and IBM Strategic Pointer 2000 Customer Module for Windows.

International TechneGroup, Inc. (ITI)
QFD Software and Services
5303 DuPont Circle
Milford, OH 45150
Phone: 800-783-9199, also 513-576-3900; Fax: 513-576-3994
email: qfdinfo@iti-oh.com

Website: http://www.iti-oh.com
Contact: Ann Grace

ITI provides training, consulting and facilitation as well as software in support of QFD and related customer-driven decision-making processes. ITI training in customer requirements collection, customer-driven decision making and objective target setting can help get an organization started in putting the power of QFD to work. ITI's consultants can also help maximize the benefits of the QFD process by working with an organization's teams as facilitators. QFD/CAPTURE is ITI's world-leading software for QFD. It puts the power of QFD to work for an organization by providing flexible data entry, analysis and presentation capabilities.

The QFD Institut Deutschland (QFD-ID)
Pohligstr. 1
50923 Cologne, Germany
Phone: 49-221-470-5369; Fax: 49-221-470-5386
email: herzwurm@informatik.uni-koeln.de
Websites: http://www.informatik.uni-koeln.de/winfo/prof.mellis/
                qfdid.htm
                http://www.informatik.uni-koeln.de/winfo/
                prof.mellis/qfdtool.htm
Contact: Georg Herswurm, Executive Director

QFD-ID provides a short list of QFD tools, including suppliers and a short comment about the advantages and disadvantages of the tools. The comments are those of the users and are not the result of an exhaustive scientific review. The list of QFD tools and comments are in German.

The QFD Institute (QFDI)
1140 Moorhead Court
Ann Arbor, MI 48103
Phone: 313-995-0847; Fax: 313-995-3810
email: qfdi@qfdi.org
Website: http://qfdi.org/www.qfdi/
Contact: Glenn Mazur, Executive Director

The QFD Institute was founded in 1993 to advance QFD methodology. The Institute conducts an annual North American Symposium on QFD, and International Symposium every four years and regular forums to provide a "safehouse" where QFD professionals can meet to discuss issues of mutual interest and topics at great length in an atmosphere of support and flexible structure. *Transactions from the Symposia on QFD,* forum minutes, other published QFD case studies and QFD-related materials are available through the Institute.

Note: A summary of the transactions from 1989 through 1996 is included at the end of this appendix.

Technicomp, Inc.
1111 Chester Avenue
Cleveland, OH 44114
Phone: 800-735-4440, also 216-687-1122; Fax: 216-687-0637
Contact: Jon Catalano, V. P.

Technicomp offers a five-tape self-contained video course on QFD principles with a case study and workbook.

TRIZ Journal
Phone: 909-949-0857; Fax: 909-949-2968
email: 73763.3077@compuserve.com
Website: http://www.triz-journal.com
Contact: Ellen Domb, Publisher

The *TRIZ Journal* has been created as an on-line journal for people interested in the TRIZ methods of creativity and innovation. The journal editors believe that we will learn more faster and in more depth if we work together, sharing what we learn as we learn it. The journal welcomes all practitioners of TRIZ—experts, beginners, traditional practitioners, software-assisted TRIZ practitioners and those who are wondering what TRIZ is and what it can do. Journal contents include question-and-answer columns, announcements of seminars and conferences, technical articles and case studies.

MAJOR SOURCE OF QFD CASE STUDIES, METHODOLOGIES, AND RELATED INFORMATION: TRANSACTIONS FROM THE ANNUAL SYMPOSIUM ON QFD

Since 1989, an annual symposium that covers all aspects of QFD has been held. The following is a listing of the transactions from the nine symposiums (1989–1997). These transactions have been listed in order by year and, within each year, by their number (or letter) order in the transactions. In addition each transaction has been categorized in three ways: by industry, by application and by focus. On the disk there is a

Y-shaped matrix keyed to this listing (using year and number/letter) that allows searching on one category and linking to either of the other two.

Year	ID	Industry	Application	Focus
89	H		Tutorial	TQM/Quality Techniques
89	L		Tutorial	Implementation
89	A	Automotive	Design	Auto Subsystems
89	E	Automotive	Design	Auto Subsystems
89	F	Automotive	Design	Auto Subsystems
89	G	Automotive	Design	Auto Subsystems
89	N	Automotive	Design	Auto Subsystems
89	O	Automotive	Design	Program/Product Development
89	B	Automotive	Tutorial	VOC
89	M	Automotive	Tutorial	Implementation
89	J	Electronics	Design	Program/Product Development
89	K	Electronics	Tutorial	Implementation
89	C	Healthcare	Design	Program/Product Development
89	I	Telecommunications	Tutorial	Program/Product Development
//////	/////	////////////////////////////	/////////////////////////	///
90	B		Tutorial	TQM/Quality Techniques
90	C		Tutorial	Implementation
90	M		Tutorial	Implementation
90	O		Tutorial	Implementation
90	P		Tutorial	Implementation
90	Q		Tutorial	Implementation
90	A		Tutorial	Program/Product Development
90	R		Tutorial	TQM/Quality Techniques
90	E		Tutorial	TQM/Quality Techniques
90	G		Tutorial	TQM/Quality Techniques
90	V	Aerospace	Design	Program/Product Development
90	K	Automotive	Design	Auto Subsystems
90	L	Automotive	Manufacturing	Process Design/Improvement
90	S	Automotive	Manufacturing	Process Design/Improvement
90	T	Automotive	Quality	TQM/Quality Techniques
90	D	Consumer Products	Tutorial	TQM/Quality Techniques
90	N	Defense	Manufacturing	Process Design/Improvement
90	H	Healthcare	Design	Design Process
90	W	Healthcare	Design	Design Process
90	I	Marketing	Tutorial	VOC
90	J	Marketing	Tutorial	VOC
90	F	Software	Design	Program/Product Development
90	X	Software	Design	Program/Product Development
90	U	Telecommunications	Customer Service	Process Design/Improvement
//////	/////	////////////////////////////	/////////////////////////	///
91	I		Design	TQM/Quality Techniques
91	G		Quality	TQM/Quality Techniques
91	H		Quality	VOC
91	B		Tutorial	TQM/Quality Techniques

Year	ID	Industry	Application	Focus
91	C		Tutorial	Implementation
91	D		Tutorial	Implementation
91	F		Tutorial	Implementation
91	J		Tutorial	Implementation
91	K		Tutorial	Implementation
91	P		Tutorial	Implementation
91	Y		Tutorial	Implementation
91	A		Tutorial	Program/Product Development
91	R		Tutorial	Program/Product Development
91	E		Tutorial	TQM/Quality Techniques
91	O		Tutorial	TQM/Quality Techniques
91	Q		Tutorial	TQM/Quality Techniques
91	T		Tutorial	VOC
91	X		Tutorial	VOC
91	AA	Automotive	Design	Auto Subsystems
91	S	Automotive	Design	Auto Subsystems
91	W	Automotive	Design	Design Process
91	N	Electronics	Design	Design Process
91	V	Marketing	Planning	VOC
91	Z	Marketing	Tutorial	VOC
91	L	Software	Design	Program/Product Development
91	U	Software	Design	Program/Product Development
91	M	Telecommunications	Design	Program/Product Development
/////	/////	////////////////////////	//////////////////////	//
92	01		Tutorial	TQM/Quality Techniques
92	02		Tutorial	Implementation
92	03		Tutorial	VOC
92	04	Manufacturing	Tutorial	Program/Product Development
92	05	Healthcare	Design	Program/Product Development
92	06		Tutorial	TQM/Quality Techniques
92	07	Manufacturing	Planning	Strategic Planning
92	08	Automotive	Design	Process Design/Improvement
92	09		Purchasing	TQM/Quality Techniques
92	10	Telecommunications	Customer Service	Process Design/Improvement
92	11		Quality	TQM/Quality Techniques
92	12	Service	Tutorial	ProcessDesign/Improvement
92	13		Tutorial	Design Process
92	14		Tutorial	Design Process
92	15	Consumer Products	Design	Program/Product Development
92	16		Tutorial	TQM/Quality Techniques
92	17		Tutorial	VOC
92	18	Service	Design	Process Design/Improvement
92	19	Software	Design	Program/Product Development
92	20		Tutorial	VOC
92	21	Software	Design	Program/Product Development
92	22	Healthcare	Design	Design Process
92	23	Education	Design	Program/Product Development
92	24	Service	Design	Program/Product Development

Year	ID	Industry	Application	Focus
92	25	Defense	Quality	TQM/Quality Techniques
92	26		Tutorial	TQM/Quality Techniques
92	27	Telecommunications	Human Resources	Program/Product Development
92	28	Healthcare	Design	Program/Product Development
92	29	Utility	Design	Program/Product Development
92	30	Aerospace	Design	Design Process
92	31	Manufacturing	Planning	Strategic Planning
92	32	Consumer Products	Marketing	VOC
92	33	Consumer Products	Manufacturing	Process Design/Improvement
92	34	Automotive	Design	Auto Subsystems
92	35	Software	Design	Program/Product Development
92	36	Electronics	Human Resources	Program/Product Development
92	37	Aerospace	Design	Program/Product Development
//////	/////	////////////////////////////	////////////////////////////	///
93	01	Healthcare	Marketing	VOC
93	02		Tutorial	TQM/Quality Techniques
93	03		Tutorial	Implementation
93	04		Tutorial	VOC
93	05		Tutorial	VOC
93	06		Tutorial	Process Design/Improvement
93	07		Tutorial	Implementation
93	08		Tutorial	Design Process
93	09	Aerospace	Design	Design Process
93	10	Manufacturing	Design	Program/Product Development
93	11		Tutorial	Design Process
93	12	Aerospace	Design	
93	13	Consumer Products	Tutorial	Implementation
93	14		Human Resources	VOC
93	15		Tutorial	TQM/Quality Techniques
93	16		Tutorial	Design Process
93	17		Tutorial	TQM/Quality Techniques
93	18	Automotive	Design	Program/Product Development
93	19	Utility	Customer Service	VOC
93	20	Construction/ Building	Planning	
93	21	Defense	Planning	
93	22	Defense	Planning	
93	23	Defense	Design	
93	24	Service	Design	Process Design/Improvement
93	25	Defense	Planning	
93	26	Automotive	Design	Auto Subsystems
93	27	Automotive	Tutorial	Implementation
93	28	Automotive	Design	Program/Product Development
93	29	Automotive	Design	Auto Subsystems
93	30	Automotive	Design	Design Process

Year	ID	Industry	Application	Focus
93	31	Construction/ Building	Design	Program/Product Development
93	32	Telecommunications	Design	VOC
93	33		Quality	TQM/Quality Techniques
93	34	Healthcare	Customer Service	
93	35	Marketing	Tutorial	Program/Product Development
93	36	Consumer Products	Manufacturing	Process Design/Improvement
93	37		Human Resources	VOC
93	38	Manufacturing	Tutorial	Process Design/Improvement
93	39		Tutorial	TQM/Quality Techniques
93	40	Telecommunications	Customer Service	Implementation
93	41	Service	Tutorial	VOC
93	42	Education	Design	Program/Product Development
93	43	Electronics	Tutorial	Implementation
93	44	Defense	Design	
//////	/////	////////////////////////////	////////////////////////	//
94	01		Tutorial	TQM/Quality Techniques
94	02		Tutorial	Strategic Planning
94	03	Electronics	Marketing	Program/Product Development
94	04		Design	Design Process
94	05	Defense	Design	
94	06		Manufacturing	Process Design/Improvement
94	07	Construction/ Building	Design	Program/Product Development
94	08	Aerospace	Tutorial	Process Design/Improvement
94	09	Electronics	Tutorial	Implementation
94	10		Tutorial	VOC
94	11		Human Resources	VOC
94	12	Marketing	Planning	VOC
94	13		Tutorial	Implementation
94	14		Tutorial	Implementation
94	15	Healthcare	Planning	Strategic Planning
94	16		Tutorial	TQM/Quality Techniques
94	17	Electronics	Manufacturing	Process Design/Improvement
94	18	Service	Tutorial	Program/Product Development
94	19	Software	Design	Program/Product Development
94	20	Software	Design	Program/Product Development
94	21		Tutorial	TQM/Quality Techniques
94	22	Education	Human Resources	TQM/Quality Techniques
94	23	Education	Planning	Program/Product Development
94	24	Education	Planning	VOC
94	25		Tutorial	Strategic Planning
94	26	Healthcare	Customer Service	VOC
94	27	Healthcare	Customer Service	VOC
94	28	Healthcare	Customer Service	VOC

Year	ID	Industry	Application	Focus
94	29	Healthcare	Tutorial	
94	30	Defense	Design	
94	31	Aerospace	Quality	TQM/Quality Techniques
94	32	Construction/ Building	Design	Program/Product Development
94	33	Electronics	Design	Program/Product Development
94	34	Automotive	Design	Auto Subsystems
94	35	Automotive	Design	Program/Product Development
94	36	Automotive	Design	Program/Product Development
94	37	Automotive	Manufacturing	Process Design/Improvement
94	38	Automotive	Design	Auto Subsystems
94	39	Manufacturing	Design	Program/Product Development
94	40		Tutorial	Implementation
94	41	Service	Design	
94	42	Defense	Planning	
//////	/////	//////////////////////////////	/////////////////////////	//
95	01		Tutorial	TQM/Quality Techniques
95	02	Construction/ Building	Design	Program/Product Development
95	03		Tutorial	TQM/Quality Techniques
95	04		Tutorial	TQM/Quality Techniques
95	05		Tutorial	Design Process
95	06		Tutorial	TQM/Quality Techniques
95	07		Tutorial	TQM/Quality Techniques
95	08		Tutorial	TQM/Quality Techniques
95	09	Healthcare	Customer Service	
95	10	Healthcare	Customer Service	
95	11		Tutorial	VOC
95	12	Utility	Customer Service	
95	13	Education	Planning	TQM/Quality Techniques
95	14	Education	Planning	TQM/Quality Techniques
95	15		Tutorial	VOC
95	16		Tutorial	TQM/Quality Techniques
95	17		Tutorial	Strategic Planning
95	18		Tutorial	Strategic Planning
95	19		Tutorial	TQM/Quality Techniques
95	20		Tutorial	TQM/Quality Techniques
95	21		Tutorial	TQM/Quality Techniques
95	22	Utility	Design	
95	23	Automotive	Design	Auto Subsystems
95	24	Consumer Products	Tutorial	Implementation
95	25	Construction/ Building	Design	Design Process
95	26	Service	Customer Service	Implementation
95	27	Telecommunications	Human Resources	
95	28	Telecommunications	Human Resources	Program/Product Development

Year	ID	Industry	Application	Focus
95	29		Human Resources	VOC
95	30	Telecommunications	Manufacturing	Program/Product Development
95	31	Software	Design	Program/Product Development
95	32	Software	Design	Program/Product Development
95	33	Automotive	Quality	TQM/Quality Techniques
95	34	Utility	Quality	TQM/Quality Techniques
95	35	Education	Planning	Program/Product Development
95	36		Tutorial	TQM/Quality Techniques
95	37	Construction/ Building	Tutorial	
95	38	Telecommunications	Quality	TQM/Quality Techniques
95	39	Automotive	Tutorial	
95	40	Software	Customer Service	Program/Product Development
95	41		Tutorial	Design Process
95	42	Software	Design	Program/Product Development
95	43	Software	Design	Program/Product Development
95	44		Tutorial	Process Design/Improvement
//////	/////	/////////////////////////////	///////////////////////////	//
96	01	Manufacturing	Tutorial	Implementation
96	02	Electronics	Tutorial	VOC
96	03		Tutorial	Implementation
96	04	Consumer Products	Manufacturing	Process Design/Improvement
96	05	Construction/ Building	Design	Design Process
96	06	Construction/ Building	Design	Design Process
96	07	Healthcare	Customer Service	Program/Product Development
96	08	Healthcare	Planning	Program/Product Development
96	09	Service	Customer Service	Program/Product Development
96	10	Electronics	Customer Service	TQM/Quality Techniques
96	11	Construction/ Building	Planning	Strategic Planning
96	12		Planning	Strategic Planning
96	13	Software	Planning	Program/Product Development
96	14	Software	Design	TQM/Quality Techniques
96	15		Tutorial	Implementation
96	16	Software	Design	Program/Product Development
96	17		Tutorial	Implementation
96	18		Tutorial	Implementation
96	19	Consumer Products	Design	Program/Product Development
96	20	Consumer Products	Design	Program/Product Development
96	21	Construction/ Building	Design	Design Process
96	22		Tutorial	Strategic Planning
96	23	Electronics	Design	Design Process
96	24		Tutorial	Program/Product Development
96	25		Tutorial	TQM/Quality Techniques

Year	ID	Industry	Application	Focus
96	26		Tutorial	TQM/Quality Techniques
96	27	Electronics	Design	
96	28	Healthcare	Design	Program/Product Development
96	29		Human Resources	Design Process
96	30		Tutorial	TQM/Quality Techniques
96	31		Tutorial	TQM/Quality Techniques
96	32		Tutorial	TQM/Quality Techniques
96	33		Tutorial	TRIZ
96	34		Tutorial	TRIZ
96	35		Tutorial	TRIZ
96	36		Tutorial	TRIZ
96	37		Tutorial	TQM/Quality Techniques
96	38		Tutorial	TQM/Quality Techniques
96	39		Tutorial	TQM/Quality Techniques
96	40	Service	Tutorial	Process Design/Improvement
96	41		Tutorial	TQM/Quality Techniques
//////	/////	//////////////////////////////	///////////////////////////	///
97	01	Manufacturing	Tutorial	Implementation
97	02	Automotive	Design	Program/Product Development
97	03	Manufacturing	Tutorial	Implementation
97	04	Electronics	Design	Program/Product Development
97	05	Healthcare	Design	Program/Product Development
97	06	Automotive	Design	Program/Product Development
97	07		Quality	VOC
97	08	Telecommunications	Marketing	Program/Product Development
97	09	Healthcare	Marketing	Program/Product Development
97	10	Telecommunications	Tutorial	Design Process
97	11		Tutorial	TQM/Quality Techniques
97	12		Tutorial	Implementation
97	13		Human Resources	Process Design/Improvement
97	14	Education	Human Resources	Program/Product Development
97	15		Planning	Strategic Planning
97	16		Planning	Program/Product Development
97	17	Software	Design	Design Process
97	18		Design	Program/Product Development
97	19	Automotive	Planning	VOC
97	20		Tutorial	Implementation
97	21		Tutorial	Implementation
97	22		Tutorial	TRIZ
97	23		Design	TRIZ
97	24		Design	TRIZ
97	25		Tutorial	TRIZ
97	26		Tutorial	TRIZ
97	27		Planning	Strategic Planning
97	28		Planning	TQM/Quality Techniques

Year	ID	Industry	Application	Focus
97	29	Defense	Quality	TQM/Quality Techniques
97	30		Design	Strategic Planning
97	31	Manufacturing	Planning	Strategic Planning
97	32		Design	Design Process
97	33		Tutorial	Implementation
97	34	Automotive	Design	VOC
97	35		Tutorial	Implementation
97	36	Construction	Quality	Implementation
97	P7	Telecommunications	Customer Service	Implementation

QFD SYMPOSIUM TRANSACTIONS: 1989-1996

Y-MATRIX: INDUSTRY vs. APPLICATION
INDUSTRY vs. FOCUS
APPLICATION vs. FOCUS

345

APPENDIX C

QFD PLANNING AND REVIEW FORMS

The New Product and Service Development Plan

Name of the Product/Service Concept :_____

Product/Service Concept Team Leader: _____

Date of Presentation: _____

Revision Number: _____

Source: The Executives' Handbook on QFD. J. Moran and J. ReVelle, Markon, Inc, 1994

Section 1: Concept Overview

1. Concept description:

2. Core assumptions:

3. Issues to be researched:

Issues	Research Methodology	Findings
•		
•		
•		
•		
•		

Source: The Executive's Handbook on QFD: Defining Managements' Roles and Responsibilities

Section 2: Market Analysis

1. Best estimate of the date of market entry:

    Quarter: _____   Year: _____

    A. Issues that would delay the entry date and their impact on that
        entry date.

    Issues   Length Of Delay If Unresolved

    •

    •

    •

    •

    •

    •

    •

    B. Is there a market window? If yes, describe it.

Source: The Executive's Handbook on QFD: Defining Managements' Roles and Responsibilities

Section 2: Market Analysis (cont.)

2. Market scope:

   A. Documentation for estimates:

   B. Market size:

	Domestic	International
1. Now	$$_____ and Units _____	$$_____ and Units_____
2. 5 Years	$$_____ and Units _____	$$_____ and Units_____
3. 10 Years	$$_____ and Units _____	$$_____ and Units_____

   C. Market target location list with dollar and volume projections:

Target Location	Domestic/International	$$ Potential	Volume Potential

   -
   -
   -
   -
   -

   D. How will this market share be achieved? What will be the approach utilized?

Source: The Executive's Handbook on QFD: Defining Managements' Roles and Responsibilities

---

Section 2: Market Analysis (cont.)

3. Pricing strategy:

Source: The Executive's Handbook on QFD: Defining Managements' Roles and Responsibilities

Section 3: Competitive Analysis

1. Competitive position in target markets in $$ and units: now, and in five and ten years.

Competition	Now $$/Units	5 Years $$ Units	10 Years $$/Units
•			
•			
•			
•			
•			

2. Anticipated competitive reaction:

Competitor	Reaction
•	
•	
•	
•	
•	

3. Is the competition working on this concept now? If yes, what is their anticipated market entry date?

Competitor	Entry Date
•	
•	
•	
•	
•	

Source: The Executive's Handbook on QFD: Defining Managements' Roles and Responsibilities

Section 4: Organizational Impact

1. Value and risk analysis:

   A. Describe the strategic value to the corporation:

   B. Risks to the corporation concerning pursuit and non-pursuit of this concept:

Trend	Positive/Negative	Probability of Occurrence
•		
•		
•		
•		
•		

   C. Overall probability of success of this concept in the marketplace and the assumptions to support it:

Source: The Executive's Handbook on QFD: Defining Managements' Roles and Responsibilities

Section 4: Organizational Impact (cont.)

2. Organizational impact:

   <u>Need</u>   <u>Required</u>   <u>Cost</u>

   •

   •

   •

   •

   •

3. Regulatory/Environmental requirements and impacts:

4. Patent position:

   A. Results and data of current patent search:

   B. Potential of exclusivity:

Source: The Executive's Handbook on QFD: Defining Managements' Roles and Responsibilities

Section 5: Customer Analysis

1. Customer perceived value:

2. Customer requirements:

Customer Requirement	Satisfaction Target	Measurement Strategy
•		
•		
•		
•		
•		

3. Internal requirements:

Internal Requirement	Satisfaction Target	Measurement Strategy
•		
•		
•		
•		
•		

Source: The Executive's Handbook on QFD: Defining Managements' Roles and Responsibilities

Section 6: Financial Justification

1. Financial analysis:

2. Feasibility and development preliminary budget:

Source: The Executive's Handbook on QFD: Defining Managements' Roles and Responsibilities

Section 7: Recommendations and Conditions

1. Recommendation:

2. Approval/Conditions:

3. Review dates:

Source: The Executive's Handbook on QFD: Defining Managements' Roles and Responsibilities

APPENDIX D

QFD AND THE EXPANDED KANO MODEL

The initial Kano Model defined three types of quality—(1) Expected, (2) Normal and (3) Exciting—and presented them in a two-dimensional space (i.e., two axes, vertical and horizontal). The vertical axis shows the extent that the customer is satisfied and the horizontal axis shows the extent to which that particular quality is actually deployed (see Figure D1). It is an excellent approach to categorizing and presenting the concepts that any organization must address if it is to excel. This approach has captured the imagination of many people in the years since its introduction, and they have built on Kano's foundation. An additional aspect of the Kano Model is that, over time, features and functions that were considered Exciting Quality when introduced migrate to becoming Normal Quality and, later, to Expected Quality.

The Kano Model and related concepts are very powerful because they assist any organization to understand how to get information from its various sets of customers about those customers' wants, needs, desires and requirements. This knowledge helps the organization design information-gathering processes that are both effective and efficient and, typically, ongoing. In some organizations these information systems are called marketing research and customer feedback. The Expanded Kano Model details customer perceptions of the different types of quality and gives insight on how to gather information to address them (Figure D2).

EXPECTED QUALITY

Expected Quality (also called Basic Quality) is the minimum for entry into your market. Some aspects of Expected Quality address safety, reliability and durability issues. In talking about these items, people say, "It goes without saying," that the product is safe to use (as intended), that it will operate within normal temperatures (say, between 10 and 120°F)

Understanding your Customers' Needs better than your Customers do !!!

Many people insist "Customers don't know what they want"... They're wrong! Customers do know what they want, but they are not very good at articulating it. When you understand the 3 types of customer needs & how to uncover them, you will be well on your way to understanding the customer's needs better than they do. Below is a description, based on the "Kano Model", that is very useful in thoroughly understanding your customers' needs. The information gathered here is an excellent input into a proper QFD application.

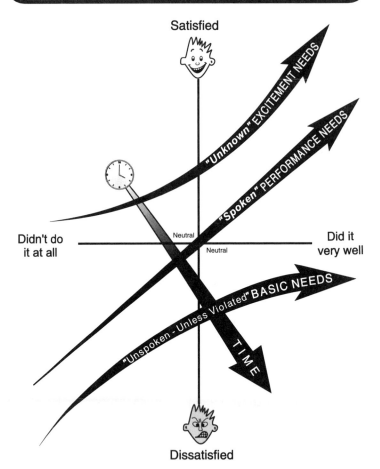

Figure D1 Kano Model.

and that it will last more than six months. These are people's assumptions about products of this type. Other aspects include assumptions that the product will interface with the surrounding world without problems or any special provisions [i.e., a device that uses electricity will use readily available, standard-sized batteries or normal house current (115 V,

"EXCITEMENT" NEEDS / INNOVATIONS !
(Make you a leader in the market)

Description:

* Customer Delights
* Wow's / Gee-Wizz's

* Thoughtful Engineering
* Innovations

* Leap Improvements
* Neat Stuff

Strategies / Sources:

* Watch customers, look for:
  - Frustrations/Annoyances
  - Customer Modifications
  - Wierd uses of product
  - Time consuming activities

* "Sales Points" in QFD
* Conflicts in QFD "Roof"
* Lateral Benchmarking
* Adding new functions by
  expanding your "scope"
  (Up/Downstream events)

* Involve Outsiders
* Talk to "Early Adopters"
* Technological Forcasting
  (TRIZ /TIPS)
* Trends that may impact the
  evolution of your product

PERFORMANCE NEEDS
(Keep you in the market)

Description:

* Voice of Customer
* Spoken Wants / Needs
* Customers' Requirements

* More the better type characteristics
* What you consciously think about
  and evaluate when you buy

Strategies / Sources:

* "Classic" Market Research
* Focus Groups
* Surveys/Clinics
* SOW/Req's Document

* Customer Interviews
* Existing Company
  information
* Contextual Inquiry

BASIC NEEDS
(Get you in the market)

Description:

* Customer Assumptions
* Things you don't think about
  unless you dont have them.

* Expected Qualities
* Expected Functions
* Things that "go without saying"

Strategies / Sources:

* Your Experience
* Customer Complaints
* Industry Standards

* Observe what all your
  competitors are doing
* Detailed Function Trees

Illustration by D.M. Verduyn

Figure D2   Expansions on the Kano Model.

60 Hz) and will come with normal cords and plugs]. Products such as video cassette recorders (VCRs) to be used in the United States are assumed to use the VHS standard (not Beta) in the NTSC format (not PAL or SECAM). There are a variety of standards for all types of industries. If you are offering a product or service for sale in a certain geographic area or industry, it is assumed that it meets the standards that prevail in that area and industry. If it does not meet those standards and the customer is surprised by the discovery, there will be customer dissatisfaction and difficulty in marketing.

Because Expected Quality "goes without saying," it takes special planning to gather the customers' expectations about Expected Quality items. A normal survey, where a set of standard questions are asked and a person responds, whether written or oral, is not going to come close to capturing all the issues involved in Expected Quality. To be able to gather the most information and understand the Expected Quality items, it is necessary to have a lengthy discussion with multiple customers with varied backgrounds. Persons who actually use the product or service (frequent and infrequent users often have different problems, so both need to be included) plus persons responsible for providing support, such as maintenance personnel, and suppliers of consumables should be part of these discussions.

A focus group would be ideal for establishing a baseline of Expected Quality issues. The best place for holding the focus group meeting would be in a location where the product or service is in actual use. If that is not possible, at least visit the place where the product/service is in use after the focus group discussions have detailed the important issues. Very often, additional issues are surfaced by this approach.

Other sources of information about Expected Quality are reviews of customer complaints and warranty returns (with related comments), items included in the limited warranty (not all are Expected Quality, some may be Normal Quality) and all the related industry standards. In addition, there may be many cultural standards, even de facto standards that vary between different cultural communities within the same geographic region. The use of color and packaging can influence purchases or the propensity to purchase. Some markets are sensitive to sales of cosmetics that have been tested on animals. In certain markets some ingredients are not acceptable to large segments of the population. Typically, soap is made from beef tallow, but in India soap is made from vegetable oils such as palm oil and rice bran oil. In Germany it is assumed you will design with recycling in mind, thus affecting even large products such as automobiles. Another source of information about Expected Quality is to observe and understand what your competitors are doing.

NORMAL QUALITY

Normal Quality (also called Performance Quality) is composed of the items typically mentioned in a normal conversation about the product's or service's quality. It is what the average person discusses with his or her contemporaries when the issue of quality comes up. It is what commercials on TV are about as well as the items mentioned in ads in magazines and newspapers. Normal Quality is important because understanding it and working on the related issues is the minimum requirement for staying in the market and being successful. Normal Quality is definitely the easiest of the three qualities to gather information about.

To understand the differences between Normal Quality and Expected Quality, we look at examples in the automotive and hospitality industries. Tires are important in the operation of an automobile, and almost everyone who owns a car will have to purchase tires at some point in time. In the case of a product like tires, the Normal Quality issues would be the price and the length of a tire's warranty (how many miles it is expected or warranteed to last). However, there are additional quality issues associated with tires that are encompassed in the following questions:

Is the tire safe at speeds up to 85 mph?
Are the tire's sidewalls likely to blow out?
Will the tire fit securely on my rims?
Will the tire's laminations separate?
Will the tire tread "chunk"?
Will the tire be difficult (impossible) to balance or keep balanced?

These issues are all Expected Quality issues, and unless you have been so unlucky as to have the conditions mentioned happen to you, it is unlikely that you will talk to your contemporaries about them.

People who travel on business or pleasure have the opportunity to evaluate hotels, including their rooms and bathrooms, in several ways. Normal Quality issues would include how long it takes to check out, the size of the hotel room, existence of a refrigerator and the view from the window. The fact that the bathroom had towels, toilet paper, soap and a small bottle of shampoo would not be worthy of mention. The existence of these items are Expected Quality. But a bathroom without towels or soap (or something else) would certainly draw a comment and a call to the front desk to request housekeeping to correct the situation. Expected Quality is supposed to be there and is not worthy of comment if it is. But where the Expected Quality is not forthcoming, the customer will speak up.

There are several approaches to establishing the Normal Quality issues and accumulating information on them. Surveys, written and oral, are effective for both creating a list of Normal Quality items and ranking their importance: actions that are necessary before a design team can begin work on the House of Quality. Focus groups are much more difficult to arrange and more costly than surveys. Therefore, if a focus group is set up to address Expected Quality issues, consideration should be given to gathering information on Normal Quality issues as well. Besides surveys and focus groups, Normal Quality issues can be studied using one-on-one interviews, market research and observing persons using the product or service.

EXCITING QUALITY

Exciting Quality is embodied in product or service features or functions that the average customer would get excited about. He or she would not necessarily be a good source of information about what new features or functions should be included in a product or service. Sometimes customers, especially progressive customers, can cite certain benefits they would like to see. How those benefits would be delivered, that is, the necessary features and functions, however, usually escape them. The problem is that most customers are not aware of opportunities to enhance products or services. They are users, not designers. They do not have a technical background so they often do not know what is possible. (Nor should they need a technical background. It is not necessary to have the engineering skills to design an automobile transmission to be able to drive and shift the gears in a car. However, a transmission engineer could ask a driver if it would be useful to be able to push a button and change easily from a high-performance mode to a high-economy mode, thus getting input from a customer.)

An Exciting Quality item is often one of those "I didn't know I needed it until I tried it!" situations. For organizations that want to be considered world class, providing Exciting Quality is essential to getting in the lead and staying in the lead in their market. Getting customers to identify Exciting Quality opportunities is much more difficult than gathering information about Expected Quality. With Expected Quality customers have become used to a feature always being there (or that it has been there in the past!) and thus know if it is missing. In the case of Exciting Quality, a customer may have never experienced it, so it is far more elusive.

There are several approaches to generating ideas or concepts for Exciting Quality. Look at current customers' operations and look for customers that have modified the product or use the service in a novel way. Look for opportunities to include more operations into the same device (look upstream or downstream in the customer's process for tasks that can be included). (For example, as this is being written, Hewlett-Packard has begun to market a combination printer, fax, copier and scanner called the HP Officejet, model 330). Identify and work with progressive customers to understand their frustrations with current features and functions and note the portion of the total cycle time taken up by non-value-added, difficult or time-consuming activities.

Although it may be expensive because of the time required and the number of people involved, a focus group that meets for a day and includes five to eight progressive customers and a similar number of open-minded marketing and technical experts can raise many Exciting Quality issues. To increase its effectiveness, gather the focus group in a room that has all sorts of visuals of past, current and proposed products/services, with associated flow charts, prototypes, cutaways and mock-ups. Use both large-group and small-group formats to discuss, exchange and build on each others' ideas and concepts. Assign and rotate different marketing/technical persons to each customer for breaks, lunch and dinner and arrange for several scribes to write, sketch and capture the overall group's and the smaller groups' discussions.

The focus group should start with a discussion of the overall industry, the trends that have already been identified and as much detail as possible on what benefits customers are currently getting or asking for. Ask the customers (for all the benefits identified) what benefits are most important to them and link the answers to details about each customer's applications. Then determine which benefits are becoming assumed (moving toward being Expected Quality) and which are still fresh and considered Normal Quality. With that foundation, move on to determining the benefits that customers feel are essential in the future [ask the question that Joel Barker, author of *Future Edge* asks: "What is impossible to do today in your business (or home or whatever), but if it could be done, would fundamentally change it?"] Seek clarification as different benefits are proposed and encourage marketing and technical personnel to suggest "what if" with the customers.

APPENDIX E

SHORT HISTORY OF QFD IN JAPAN

BACKGROUND

Since the late 1960s, QFD has had broad application in new product development in Japanese industry. This has resulted in shorter product development cycle times and smoother product development processes that have yielded higher initial quality products with better defined processes needing fewer changes after production start-up and larger market share (due to earlier to market with higher quality product more closely matching user desires). Beginning in the early 1980s, the QFD product development approach has been slowly adopted by American industry, starting with the automobile industry, with assistance from the American Supplier Institute (ASI), GOAL/QPC, and the Kaizen Institute.

A major emphasis of the QFD methodology is to assure that customers will accept and even embrace a new product or service before it has been produced and brought to market. Many organizations have made changes to QFD and then built on its fundamental concepts to enhance and broaden its usefulness in the product/service development process. Innovation is necessary to increase the scope and usage of QFD, and to assist this process, it should be useful to understand QFD's origins and basis.

After World War II, virtually all of Japan's industry was decimated, and the Japanese began rebuilding with the help of Americans like Homer Sarasohn and Charles Proxman. The reconstruction efforts addressed all sectors of Japanese industry and society and were headed by General Douglas MacArthur. An early emphasis was placed on communications with the civilian population; thus the Civil Communications Section (CCS) was established, with Homer Sarasohn as the Industry Branch Chief. The CCS's goal was to set up a domestic communications system, including the manufacturing of radios.

Sarasohn, a subject matter expert (SME) from MIT, began his efforts in 1946, and 18 months later he realized that in order to continue to advance, Japanese industry would need exposure to modern management and quality techniques. His request to approve a mini-MBA type of program of instruction in these techniques was approved by General MacArthur in August 1949. A month later the first presentation was made in Tokyo. The second presentation was made in January 1950 in Osaka. The initial programs were broad based and directed at senior executives in Japanese industry. Some of the participants included Kaoru Ishikawa and Eizaburo Nishibori. There was need for a program with more specifics directed at middle managers. Sarasohn began development of such a program and contacted Bell Labs' statistician, Walter Shewhart, about coming to Japan. However, Shewhart was ill. Thus a search was made for an alternate, and Deming, whom Shewhart knew from his experience on the War Production Board, was chosen.

At this time Deming was a professor at Columbia University and was having difficulty in generating interest by American executives in applying quality techniques. Sarasohn assured him that he would be listened to in Japan. Deming arrived in Japan in midsummer 1950, about the same time North Korea invaded South Korea. This event resulted in America refocusing its efforts in the region Deming and Joseph M. Juran (starting in 1954) taught modern concepts of quality assurance and related tools. These techniques were absorbed and applied throughout Japanese industry, particularly in the chemical processing and manufacturing industries.

From the early 1950s to the 1960s quality had a Statistical Quality Control (SQC) emphasis. Juran promoted the integration of quality aspects into the general practice of business. This idea of integrating quality and management was seconded by Feigenbaum's (1961) book *Total Quality Control*. Ishikawa was also a proponent of TQC and, through a combination of factors, by 1965 the emphasis was on TQC.

In the 1960s Japan's automobile industry began expanding with a wider variety of vehicles and more frequent model changes. The previous emphasis on quality control in automotive manufacturing processes was surpassed by the need for quality control activities in the design process. As a result, the traditional quality control techniques used in the manufacturing processes began to migrate throughout the company [leading to Ishikawa's term Company-Wide Quality Control (CWQC)]. This companywide approach fostered techniques such as policy deployment, cross-functional management and network of control points.

The network of control points allows all levels and functions of an organization to monitor and control production through a network of linked control points and checkpoints. Each level of operations and management has its own level of detail, which is linked to the levels above and below it (see Figure E1). A network can be created for each of the important aspects of a company's business, such as quality, cost and delivery. Check points are located upstream from their corresponding control point(s). There are often multiple points where the process operators collect, organize and analyze data in a focused effort to ascertain the status of a process (i.e., is it both statistically controlled and design capable?). Performance metrics are used as measures of activity. These might include conveyor speeds, mixing times, machining speeds and feeds as well as material dimensions, hardness, densities and pH.

Located downstream from their corresponding checkpoints, control points are the points where supervisors and managers monitor the results for indications of changes in their process(es). Performance metrics determine current measures of activity, which could include defect rates (e.g., defects per million defect opportunities), process status ratios (e.g., Cp, the process capability index, and Cpk, the process performance index) and sigma levels (e.g., Motorola's Six Sigma program). Note that the control points for one level are the checkpoints for the next level. For example, the checkpoints for an operator are control points for a supervisor. Figure E1 shows the network of linkages that insure consistent, uniform and predictable process outputs. Effectively, checkpoints monitor process inputs using SQC, while control points monitor process outputs using Statistical Process Control (SPC).

DEVELOPMENT OF QFD

There were two prime reasons for the early development of QFD. The first was to create a method that would assist in the process of designing new products of assured quality. The second was the felt need to have a quality control process chart ready to assist manufacturing personnel before the initial production run occurred.

A quality control process chart details (for the manufacturing supervisors and operators on the floor) what to check during each operation, who will check it and how it should be checked, the specification range to compare with the results of the checking, as well as what action should be taken, and by whom, if the operation falls outside the specified range. Such charts had been used in Japan before 1960, but at that time they were usually prepared based on the results of volume production runs.

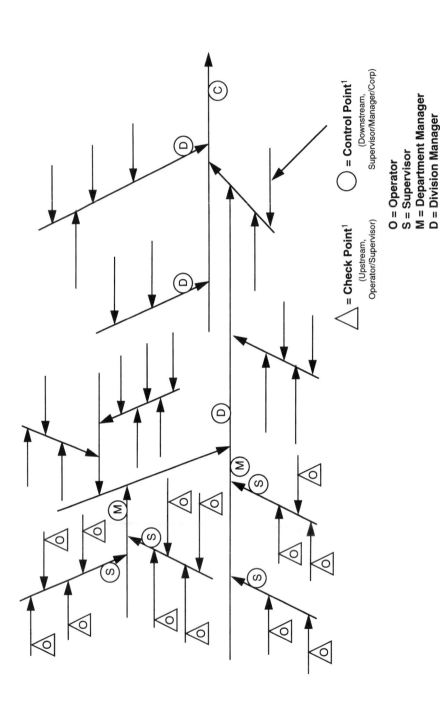

Figure E1 Network of Control and Check Points

△ = **Check Point**[1]
(Upstream,
Operator/Supervisor)

◯ = **Control Point**[1]
(Downstream,
Supervisor/Manager/Corp)

O = **Operator**
S = **Supervisor**
M = **Department Manager**
D = **Division Manager**
C = **Corporate**

[1]see definition in App. B: Glossary

Oshiumi of Bridgestone Tire contributed an idea on how to accomplish this. He created a Table of Process Assurance Items (see Figure E2, the elements inside the bold lined area), which was derived from the items of True Product Quality. [The True Product Quality items in Figure E2 are riding comfort (vibration and stiffness of ride), braking ease, ease of driving control and rolling drag.] These items are more general statements of quality rather than directly measurable ones. They are similar to statements made by customers who are asked about what qualities they need or want in a product. Because it is difficult to deliver a satisfactory item without specific measures for each quality, it is often necessary to break the True Product Quality items down into surrogate quality characteristics such as dimensions, weight, pH, hardness and adhesion, all of which can be easily measured.

The True Product Quality items were arranged in a way that showed their relationship to the more easily measured items, that is, the surrogate characteristics. The Oshiumi chart could be considered the predecessor to the House of Quality, which was developed later and showed the same relationships and information. For each surrogate characteristic, a list of process input parameters to be monitored was created for each step in the process, and then these relationships were displayed using the format of a cause-and-effect diagram. This created a network of control points that linked True Product Quality items to the parameters and parameter target values that would deliver those quality items. The cause-and-effect diagrams showed how the parameters were monitored throughout the process on a step-by-step basis with linked check and control points (see Figure E1).

Bridgestone Tire also used the Table of Product Assurance Items to review the quality assurance preparations for the manufacturing processes used in volume production. Yoji Akao liked the approach as applied to manufacturing and adapted it for use upstream in the product cycle, that is, in the new product development process. He did this by adding the elements shown outside the bold lined area in Figure E2. By connecting the items inside the bold lines, the Table of Product Assurance Items, with those outside, Akao caused the initial product quality items listed to be addressed again when the prototype and pilot production runs were complete. If, for example, a quality control process table was prepared for the rubber preparation process (done just prior to molding) and there was an interest in the quality characteristic "molded weight" of the final product, then control points such as "temperature," "specific gravity" and "blendability" of the rubber compounds being used as well as "speed (rpm)" of the mixer blades would be important.

Using the Table of Product Assurance Items with Akao's modifications, it was possible to detail the quality control points necessary to

Bold lined box contains elements later incorporated in quality table ("House of Quality")

											Product assurance items (general) specifications	Product assurance items (product) specifications	Riding Comfort	Braking Ease	Driving Control	Rolling drag
											Molded weight	Product weight	o	o	o	o

Development Process — Detailed quality design ► Process design ► Trial manufacture — Trial mass production — Initial operation — Quality design — Quality planning — Market evaluation

Quality — Detailed quality design — Process control points — Design quality (basic performance) — Planning quality — Quality requirements (current and potential) — Target quality

Product quality — Quality items assured to customers

Listing of major Q. A. items

Quality design points: Quality design — Research Improvements

Process assurance items:

Material procurement	Relin-ing	Weav-ing	Coat-ing	Band-ing	Bead-ing	Form-ing	Mold-ing	Inspec-tion
	Specific gravity of formulation rubber		Unit weight of coated cord	Band weight	Insulation bead weight		Appearance of mold	
			Unit weight of coated canvas		Tripping bead weight	Chafer weight	Mold dimensions	

Figure E2 Integrated Product and Process Development Process (with Product Assurance Items inside the bold lines)

deliver the True Product Quality items the customer required. Most importantly, this new chart and information could be provided to the floor operators before the start of volume production. Akao called this new approach Objective Quality Deployment. The approach was first tried out at Komatsu, then at Isuzu Motors and then at Fuji Motors. The results were published in the *ICQC—1969 Tokyo Proceedings* (available in English). Additional trials were performed and the results published, but little interest (and few references) were noted in subsequent literature.

In 1971, Konica Camera applied the new method to the initial volume production of Japan's first dry copier, where it was very successful. This project was reported in the April 1972 issue of *Standardization and Quality Control* magazine. In that article, Akao (1972) used the term Quality Deployment to describe the method. However, there were more developments and improvements to come.

Key among them was the quality table mentioned in the article "Ship Design and Quality Table" in the May 1972 edition of the Japanese Union of Scientists and Engineers (JUSE) *Quality Control,* written by Koichi Nishimura (1972) of Mitsubishi Heavy Industries. In 1973, another article, "QC Activities of Products Ordered in Our Company—Quality Table," in *Quality Control,* written by Akira Takayanagi, described the quality table as one "showing the correlation between the True Product Quality (i.e., what the customer wants), as represented by a set of systematically organized functions, and the surrogate characteristics that can be referred to in place of the True Product Quality characteristics." Shigeru Mizuno and Yasushi Furukawa assisted in the creation of the quality table.

Creating the quality table begins by collecting the customer's quality requirements. The goal is to organize and carry the identified requirements as far as possible downstream in the production process to assure the customer's requirements are reflected in the final product. In 1975, a QFD study group was created by the Japanese Society for Quality Control with Akao as chairman. By 1978, there were enough case studies of QFD that Mizuno and Akao edited a book published by JUSE, *Quality Function Deployment: An Approach to Company-Wide Quality Control.* At this point, the most advanced QFD technique was practiced at Kubota Steel with the assistance of Furukawa and others.

Hino Motors (Toyota Group) and Toyota Auto Body began using QFD in 1975. In 1979, these two companies made presentations on QFD along with a lecture by Y. Akao to Toyota Motor Company, and from that point forward QFD was quickly introduced to the remainder of the Toyota Group. In the United States, the quality table is often called the House of Quality. T. Sawada initiated this term while he was using QFD to develop the Toyota Light Ace van. Since 1982, a 3-hour lecture on QFD developed by Akao has been part of Toyota's basic quality course.

Mizuno and Akao (1978) give definitions for quality deployment, quality function deployment (narrow sense) and quality function deployment (broad sense). These can be found in the Glossary.

Between 1979 and 1983, discussions on the relationship between quality function and reliability were held. Subsequently, Osamu Furukawa proposed a plan to define the relationship of quality deployment with cost deployment and technical deployment. In 1983, these studies were combined into a quality deployment procedure. This quality control procedure was published along with case studies from chemical, construction, service and software companies in the book *Practical Applications of Quality Deployment in New Product Development* (Akao, 1988a) by the Japanese Standards Association.

In 1988, a QFD study group was formed within JUSE with Akao as its chairman. At that time, QFD seminars were held on a regular basis sponsored by various organizations. Among these seminars and sponsors were the following:

Four-day seminar	JUSE
Three-day course	JSA (Japanese Standards Association)
Two-day course	Central Japanese Quality Control Association
Two-day course	Japan Productive Science Association

Kogure and Akao (1983) introduced QFD to the United States in an article in *Quality Progress* magazine and through their first U. S. presentation: "Seminar on Company-Wide QC with Quality Deployment" (held in Chicago, Illinois, October 31–November 3, 1983). Subsequent efforts by Larry Sullivan of the American Supplier Institute and Bob King of GOAL/QPC have led to the establishment of on-going QFD seminar programs throughout the United States ever since.

For additional information on how the foundations of QFD were established see Mizuno and Akao's (1994) book, *QFD: The Customer-Driven Approach to Quality Planning and Development.* The authors, in communication with Akao, were told that the early events in building the QFD methodology are detailed in the appendix titled "Developmental History of QFD" by Mizuno and Akao (1993), a translation by Glenn Mazur of the original text, *Quality Function Deployment: An Approach to Company-wide Quality Control,* (Mizuno and Akao, 1978).

APPENDIX F
ABOUT THE DISK

DISK CONTENTS

The enclosed disk contains:

- **QFD/Pathway software**—For a detailed description of this software refer to Appendix A.
- **QFD Symposium Transactions Y-matrix**—The contents of this matrix appear as a table in Appendix B. The contents of the referenced table were reformatted as a Y-matrix and saved on this disk in Microsoft Powerpoint version 4.0.

MINIMUM SYSTEM REQUIREMENTS

QFD/Pathway Software:

- IBM PC or compatible computer
- 386 or higher processor
- Microsoft Windows 3.1, Windows 95, Windows NT (3.5.1 or higher)
- 8 MB RAM (16 MB recommended)
- 1 MB hard disk space (for QFD/Pathway software)
- 3.5" floppy disk drive

QFD Symposium Transactions Y-matrix:

To view the QFD Symposium Transactions Y-matrix chart, you need to have the above system as well as Microsoft Powerpoint (version 4.0 or higher) or a Microsoft Powerpoint viewer installed on your computer. Several versions of Microsoft Powerpoint viewers are available free from the Microsoft ftp server at ftp://ftp.microsoft.com/Softlib. For more information about downloading free Powerpoint viewers go to the Microsoft homepage at http://www.microsoft.com/powerpoint.

INSTALLATION INSTRUCTIONS FOR QFD/PATHWAY SOFTWARE

To install the QFD/Pathway software onto your computer's hard drive do the following:

1. Insert the floppy disk into the floppy disk drive.
2. Windows 3.1 or Windows NT 3.5.1: Go to Program Manager. From the File pull-down menu, choose Run.
   Window 95 or Windows NT 4.0: From the Start button, choose Run.
3. Type **A:\SETUP** and hit the Enter key.
4. The setup program screen will appear. The default destination directory is C:\QPATH. If you wish to change the destination directory you may do so now. Hit the Continue button.
5. The setup program will install the program to your hard drive.

HOW TO USE THE SOFTWARE AND FILES

To launch the QFD/Pathway program, double click on the QFD/Pathway icon from the Program Manager in Windows 3.1 or choose the icon from the Start Menu in Windows 95.

To read the QFD Symposium Transaction Y-matrix chart, you must have Microsoft Powerpoint version 4.0 or higher or have access to a Microsoft Powerpoint viewer. To view the chart included on this disk, insert the disk into your floppy disk drive and launch the Microsoft Powerpoint software. Go to the pull-down menu and choose File, Open. Type **A:\QFDSYM\QFDSYM.PPT** and press Enter.

USER ASSISTANCE

If you need basic assistance with installation, or if you have a damaged disk, please call the John Wiley & Sons product support number at (212) 850-6753, weekdays between 9 AM and 4 PM Eastern Standard Time. You can also reach us through email at techhelp@wiley.com.

If you have any questions or comments about the software and files on the disk, please contact the authors Robert A. Kemerling at rak12146@cinci.infi.net or Jack ReVelle at jbrevelle@theriver.com.

To place additional orders or to request information about other Wiley products, please call (800) 225-5945.

GLOSSARY

Acceptance matrix   A manufacturing process matrix that contains cells that represent the ability of a manufacturing product test to predict a product's performance with respect to a specific performance measure.

Activity Network Diagram (AND)   Also known as PERT (Program Evaluation and Review Technique), CPM (Critical Path Method), and arrow diagram, it assists a team to find the most efficient path and realistic schedule for the completion of any multievent project.

Affinity analysis   Also known as affinity diagram, card sort, person card sort, K-J, Shiba method and whole brain affinity, it is used by a team to help sort large volumes of data or responses (such as verbatims) into similar categories through a silent consensus process. It is perhaps the most widely used of the seven management and planning tools and is used whenever structure is necessary. This is accomplished by organizing textual information into related groups.

Algorithm   A mathematically based model used in the analysis of a QFD matrix to determine which of the substitute quality characteristics (the HOWs) are most important and should be deployed to subsequent matrices for more detailed analysis.

American Supplier Institute (ASI)   An American nonprofit organization founded in the early 1980s as the Ford Supplier Institute that focuses on QFD, robust design, TRIZ (Theory of Innovative Problem Solving, or TIPS), FMEA, and other Total Quality Management (TQM) tools (see Appendix B).

Analytical Hierarchy Process (AHP)   A formal procedure used to rank alternatives so that the resulting scores are variable in nature.

Benchmarking, competitive   This activity is usually collectively performed by the marketing members of a QFD team. It is a comparison of customer perceptions of "our" ability to satisfy customer demands (the WHATs) versus the abilities of the competition.

Benchmarking, technical   This activity is usually collectively performed by the engineering members of a QFD team. It is a performance comparison of various technical features (substitute quality characteristics, also known as the HOWs) of "our" products to those of the competition.

Brainstorm   A process used by a team or other groups of persons to generate as many ideas as possible in a given amount of time, e.g., a 1-hour meeting.

Capability Deployment   A process that identifies factors that must be successfully met in new product or service development. Charts C-1′ and C-1″ are used for this process.

Catch ball   Refers to communications activities, whether in an upward, downward or horizontal direction, that must travel back and forth between persons several times to be clearly understood.

Cause-and-effect diagram   A graphing tool used to identify substitute quality characteristics (the HOWs) from demanded quality items (the WHATs) in the QFD process. The cause-and-effect diagram is adapted to focus the QFD team on one demanded quality element at a time.

Checkpoint   Located upstream from its corresponding control point, these are multiple points where process operators collect, organize and analyze data in a focused effort to ascertain the status of a process, i.e., is it both statistically controlled and design capable? Performance metrics are used to determine current measures of activity, which could include conveyor speeds, spindle speeds, activity cycle times and oven temperatures as well as material dimensions, densities and purity levels.

Company plan   Contained in the A1 matrix of the 30-matrix (GOAL/ QPC) approach to QFD, the company plan is the decision that results from comparing customer evaluations of one company's ability to satisfy each customer demand (the WHATs) versus its competitors, i.e., competitive benchmarking.

Concept Deployment   A process to define alternatives for achieving functions. Chart C-3 is used to support this process.

Concept-level design   A process that uses both concept deployment and capability deployment to identify methods for and alternatives to achieving functions needed by the customer in a product or service.

Concept Selection (Pugh)   Developed in Scotland by Stuart Pugh as a procedure for the generation of new designs, its most frequent application is found in the use of the E-series matrices contained within the Matrix of Matrices, the 30-matrix (GOAL/QPC) approach to QFD. Evaluation of several alternative designs for a product or process determines which is the most appropriate based upon previously selected criteria.

Concept/technology   An engineering method for achieving a specific desired function. Usually identified during the concept deployment process.

Concurrent engineering   Also known as simultaneous engineering, it is the way future-focused organizations jointly plan, develop and produce their products as well as the processes they will use to produce those products.

Continuous improvement   Also know as *kaizen,* can be either an individual and/or organizational effort to incrementally and regularly improve process outputs through the judicious application of Total Quality Management (TQM) tools.

Control point   Located downstream from its corresponding checkpoints, this is the point where supervisors and managers monitor results for indications of changes in their process(es). Performance metrics are used to determine current measures of activity, which could include defect rates (e.g., defects per million defect opportunities), process status ratios (e.g., Cp = process capability index and Cpk = process performance index) and sigma levels (e.g., Motorola's Six Sigma program). *Note:* One level's control points are the next level's checkpoints. For example, an operator's control points are a supervisor's checkpoints. Thus, Figure E1 is really a network of link-pins to ensure consistent, uniform and predictable process outputs. Effectively, checkpoints monitor process inputs using Statistical Quality Control (SQC) while control points monitor process outputs using Statistical Process Control (SPC).

Core competencies/technologies   Every organization possesses certain unique capabilities that it performs with special prowess. Successful organizations usually focus their production/service efforts in those areas where they have highly competitive core competencies/ technologies.

Correlation   Refers to the degree of statistical relationship between two factors/variables, i.e., the extent to which the magnitude of one factor explains or influences the value of another factor. For example, height (taller or shorter) influences weight (greater or lesser).

Countermeasures   Proactive changes to a product or process design that are intended to eliminate or reduce the likelihood of a specific event.

Customer demands   The result of converting the Voice of the Customer into understandable, positively oriented statements to which a supplier can react and/or respond. Customer demands are sometimes referred to as the WHATs.

Customer words   Also known as verbatims, these are untranslated written or oral expressions received from surveys, audits, focus groups and other sources that describe customer desires, i.e., the attributes or qualities noted in the Kano Model.

Demanded quality   A demanded quality statement is one that is understandable and deployable by an organization. It is the result of transforming the customers' verbatims, the customer's subjective description of what functions a product should have as well as how it should perform. Ideally, customer verbatims are worded as positive statements (for ease of subsequent analysis) and are composed of an adjective followed by an infinitive, e.g., easy to hold.

Demanded Quality Deployment   A process that identifies the relationships between demanded quality items and substitute quality characteristics. It converts the degree of importance for demanded quality into the degree of importance for quality characteristics. Chart A-1 is used for this conversion process.

Design FMEA   An FMEA conducted on the parameter values of an engineering design.

Element   An item that must be present and operational within a part/mechanism to achieve the best concept/technology. Elements are identified during the element deployment process.

Element Deployment   A process that identifies items needed for achieving concepts and defines the relationships between those items and substitute quality characteristics. Chart A-4 is used for this identification process.

Element Failure Mode Deployment   A process that identifies the relationships between elements and their failure modes. Chart D-4 is used for this identification process.

Element-level design   A process that uses element deployment and element failure mode deployment to identify the items needed for achieving the best concepts/technologies.

Facilitator   A coordinator of the planning, design, execution and completion of the QFD Project. The facilitator works with a QFD team and is primarily concerned with the process of accomplishing QFD, not with matrix content.

Failure mode   The sudden or growing inability of a product or service to satisfy its intended task or function.

Failure Mode and Effect Analysis (FMEA)   A process where each potential failure mode in a product or service is identified and analyzed to determine its effect on the system. This is accomplished using an L-shaped matrix (also called FEMCA-Failure Mode, Effect and Criticality Analysis). It is a method of prioritizing the manufacturing or engineering design causes of failure. Corrective actions are developed and implemented for the most important causes.

Fault tree analysis   A graphical technique utilizing backward analysis to investigate or examine potential failures or hazards in a product, service or system. It is a structured way of demonstrating relationships between failures and their causes.

Focus group   A group of customers who use and evaluate a product or service as well as competitor's products/services in the presence of a QFD team.

Function   The characteristics and tasks necessary for a product or service to provide benefits and be acceptable to a customer.

Function (organization)   Various areas of specialization within an organization, e.g., engineering, manufacturing, marketing, finance, etc.

Function (product)   What a product is supposed to do or the task it is to perform. Ideally, it is worded as an active verb followed by an object, e.g., for a hammer the function is to "drive nails."

Function Deployment   A process that identifies the tasks necessary to meet demanded quality, as well as the relationships between functions, demanded quality items and quality characteristics. Function deployment converts the degree of importance for each demanded quality item and substitutes quality characteristic to the degree of importance for each function. Charts B-1 and A-2 are used for this conversion process.

Function Tree   A graphing tool used to identify concepts/technologies for achieving functions needed in a product or service.

Gemba   The environment within which a customer uses or consumes a product or service.

GOAL/QPC   An American nonprofit organization founded in the early 1980s as GOAL [Growth Opportunity Alliance of Lawrence

(MA)], now known as GOAL/QPC (GOAL/Quality, Productivity, and Competitiveness). Its organizational focus is on QFD, Policy Deployment (also known as *hoshin kanri* or *hoshin* planning), 7-MP (the seven management and planning tools) and other Total Quality Management (TQM) tools (see Appendix B).

Hin-shitsu ki-no ten-kai Six Japanese Kanji characters that were translated into English as Quality Function Deployment by Donald Clausing in the early 1980s.

House of Quality (HOQ) Also known as the A1 matrix (from the Matrix of Matrices), the Planning Matrix, the Requirements Matrix or the Quality Assurance. Table, it is the basic/initial QFD chart and is used to define and relate customer demands (the WHATs) to substitute quality characteristics (the HOWs).

Improvement ratio Within the A1 matrix of the 30-matrix (GOAL/QPC) approach to QFD, the value for each customer demand (the WHATs) resulting from the division of the company plan by its customers' evaluation of the company's ability to satisfy that WHAT.

Integrated Product and Process Development (IPD) A team-based implementation of concurrent engineering.

Interrelationship digraph A graphical technique used by teams to systematically identify, analyze and classify the cause-and-effect relationships that exist between all critical issues so that key drivers or outcomes can become the focus of an effective solution.

Kano Model A two-dimensional (customer reality and customer perception), graphical description of product and service characteristics that are classified as dissatisfiers (the MUST level qualities or attributes), satisfiers (the WANTs level) and delighters/exciters (the WOW level).

Kansei From the Japanese, meaning "sensory" engineering. It is used to create product designs that appeal to customers' sense of smell, touch, etc. Often involves some intangible attributes such as communicating a feeling of luxury or a special image.

K-J Refers to a comprehensive process used to analyze textual information. It is similar to affinity analysis and was developed in Japan by Jiro Kawakita.

Manufacturing FMEA An FMEA conducted on a manufacturing process.

Market planning A process that converts the Voice of the Customer into a marketing plan. The Voice of the Customer Table is used for this process.

Matrix   A rectangular array of statements of customer demands, substitute quality characteristics, functions, and so on, networked against each other so that relationships can be determined and recorded.

Matrix analysis   Also known as a matrix or matrix diagram, a graphical technique used by teams to systematically identify, analyze and rate the presence and strength of relationships between two or more sets of information. Matrices come in a variety of shapes, including L, T, Y, X and C.

Matrix Analysis Plan (MAP)   A graphical depiction of the flow of customer verbatims through the Voice of the Customer Table (VOCT) and the A1 matrix of the 30-matrix (GOAL/QPC) approach and 15 other of the more commonly used QFD charts. The MAP is used by QFD teams in their search to determine which matrices are most relevant to their project. (see Appendix A, QFD/Pathway and Matrix Interrelationships).

Matrix of Matrices   A set of 30 matrices and charts (the GOAL/QPC approach) that represent the deployment of quality throughout the disciplines of cost, reliability and concepts/technologies while maintaining the primary focus of satisfying the Voice of the Customer.

Normalized scores   Also known as relative frequency, it is a percent or fraction of the total of all scores.

Part/Mechanism   A mechanical, chemical or electrical component that must be present and operational within a product to achieve the best concept and/or technology.

PDCA (Plan, Do, Check, Act)   Originally referred to as the Shewart Cycle and more recently as the Deming Cycle, a proactive control process that is a fundamental component of QFD in that its use helps teams to avoid potential problems.

Policy Deployment   Also known as *hoshin kanri, hoshin* planning, *hoshin,* management by planning (MBP) [in contrast to management by objectives (MBO)] or management by policy (MBP), a detailed methodology used to ensure that an organization's strategic plan is fully deployed, operational and on track.

Prioritization matrices   A graphical tool used to narrow down options through a systematic approach compares choices by selecting, weighing and applying criteria.

Process Decision Program Chart (PDPC)   Based on the tree diagram, this tool is used by teams to systematically improve goal implementation through contingency planning. This graphic contains the critical processes and tasks of the design process. Columns present an organization's functional units, rectangles identify responsibility for various activities and arrows indicate the flow of documents and decisions.

Product Planning Table   One of the component parts of the Demanded Quality vs. Performance Measures Matrix. It provides for calculation of the relative importance of the performance measures and the setting of target values.

Pugh Concept Selection   See Concept Selection (Pugh)

QFD Institute   A North American nonprofit organization, founded in the early 1990s as an outgrowth of several QFD-oriented experiments. In 1994, the Institute established the first North American QFD Master Class in Dearborn, MI.

Quality Deployment   It is necessary to convert user quality requirements into counterpart technical characteristics so as to determine design quality for the finished product. Then, based on the counterpart characteristics, we systematically deploy the correlations among the quality of each functional component as well as the individual parts and each of the process elements.

Quality Function Deployment   A structured and disciplined process that provides a means to identify and carry the Voice of the Customer through each stage of product or service development and implementation. This process can be deployed horizontally through marketing, product planning, engineering, manufacturing, service and all other departments in an organization involved in product or service development.

Quality Function Deployment (narrow sense of the term)   To deploy, in detail, the jobs or business functions concerned with building up quality in end-means systems by steps (deployment of quality-assurance-oriented business functions) [per Akao (1990a)].

Quality Function Deployment (broad sense of the term)   Quality Function Deployment (narrow sense of the term) and Quality Deployment combined [per Akao (1990a)].

Quality Planning Table   One of the component parts of the Demanded Quality vs. Performance Measures Matrix. It provides for the comparison of an organization's product vs. that of the competition so as to establish the composite importance for each demanded quality.

Relationship matrix   One of the component parts of the Demanded Quality vs. Performance Measures Matrix. It provides an opportunity to determine the strength of each performance measure's ability to predict the customers' satisfaction with each demanded quality.

Relative frequency   The ratio of the frequency count of a specific event to the total number of occurrences of all the events.

Reliability   The likelihood that a product or service will satisfy its intended purpose for a specified time period.

Reliability Deployment   A process that describes failure modes for a product or service based upon demanded quality or function and identifies the associated relationships to help identify failure modes to be carried forward for further analysis. Charts D-1 and D-2 are used for this description and identification process.

Reworded data   A process of extracting all possible demands that a customer may include in a statement of wants and needs. The multiple demands contained in the customer's words are singularized via the rewording activity. This process also includes the analysis of usage.

Risk, financial   Each substitute quality characteristic (the HOWs) within the House of Quality (the A1 matrix) has a certain degree of *financial* risk associated with the organization's capability to achieve the target value established for each HOW.

Risk, schedule   Each substitute quality characteristic (the HOWs) within the House of Quality (the A1 matrix) has a certain degree of *schedule* risk associated with the organization's capability to achieve the target value established for each HOW.

Risk, technical   Each substitute quality characteristic (the HOWs) within the House of Quality (the A1 matrix) has a certain degree of *technical* risk associated with the organization's capability to achieve the target value established for each HOW.

Roof (of the House of Quality)   Also known as the A3 matrix (within the GOAL/QPC Matrix of Matrices) or as the correlation/conflict matrix, it is used to compare each substitute quality characteristic (the HOWs) to all of the others on a pairwise basis to ascertain the presence of positive and negative interrelationships, i.e., the HOWs vs. the HOWs.

Sales point   When it is recognized that a particular customer demand (the WHATs) has the necessary potential to be a delighter/exciter (the WOW level qualities or attributes), as described in the Kano Model, it is so noted in the competitive benchmarking portion of the House of Quality (the A1 matrix of the GOAL/QPC 30-matrix approach to QFD). It is used in the creation of the absolute weight.

Software, IBM Strategic Pointer   A commercial software package distributed by IBM and designed for both the PC-Windows and OS-2 environments. The *Customer* module creates, edits and manipulates the QFD House of Quality for *design* QFD (to develop new products and services) and *dynamic* QFD (to optimize existing products and services). This package also contains *SQC* and *Supplier* modules (see Appendix B).

Software, QFD/Capture   A commercial software package distributed by International TechneGroup Inc. (ITI) and designed for both the PC-Windows and Macintosh environments. QFD/Capture is used to facilitate the construction and analysis of a QFD matrix using either the 30-matrix (GOAL/QPC) or the 4-matrix (American Supplier Institute) approach. Also available is an entry-level data gathering/organizing software package, QFD/Guide (see Appendix B).

Software, QFD/Designer   A commercial software package distributed by the American Supplier Institute (ASI). QFD/Designer is designed for the PC-Windows environment only and is used to facilitate the construction and analysis of a QFD matrix using only the four-matrix (ASI) approach (see Appendix B).

Software, QFD/Pathway   A commercial software package distributed by John Wiley & Sons, Inc., it is designed for the PC-Windows environment only. QFD/Pathway is used to assist in planning a QFD project by providing a single, multimatrix pathway through the 30-matrix (GOAL/QPC) approach from the initial to the final matrix of the project. A copy of this software is provided with this text (see Chapter 5 and Appendix A).

Software Quality Deployment (SQD)   The application of QFD to the development of software products.

Specification   A detailed description of the level of performance for a component, subsystem or system that is necessary for satisfactory performance of a product.

Stakeholder   Any person who can influence a decision to use or buy a product as well as anyone who is impacted by the use of a product. Alternatively, any person who is associated with or has influence on a process.

Strategic planning   An organization's plan to achieve its mission and vision through the scheduled application of its critical/limited resources.

Substitute quality characteristic   A measurable item used to determine if the customer demands (the WHATs) will or can be achieved. Substitute quality characteristics or counterpart engineering/technical characteristics are frequently referred to as the HOWs.

System-level design   A process that uses Demanded Quality Deployment, Function Deployment, Reliability Deployment, and a portion of Capability Deployment to determine the concept/technology of a product or service.

Target (value)   A statement or value that describes, either quantitatively or qualitatively, the specified goal for a Substitute Quality Characteristic (SQC) or a function to meet stated customer needs. This is not an engineering specification. Rather, it is a statement of the desired performance level and may be better or worse than the finally selected specification. The more difficult a target is to reach, the greater the risk associated with that SQC or function.

Tree Diagram   A graphic tool used to view an objective in increasingly smaller levels of detail starting from the macrolevel and ending in the microlevel. It is an aid used to break down a complex statement or objective into its logical subparts. It can be used to break any broad goal into increasingly smaller detailed actions that must or could be done to achieve the stated goal. This tool is also referred to as the outline.

Value   (see *Target*)

Value engineering   An engineering methodology that provides cost-effective direction for a search for less expensive, less complex designs such that customer satisfaction equals or exceeds that of the original design.

Verbatims   These are the needs and wants of the customers expressed in their own exact words.

Voice of the Customer (VOC)   A spoken or written statement of the customers' wants and needs expressed in their own words. These expressed needs and wants are referred to as verbatims.

Voice of the Customer Table (VOCT)   A chart used to organize, translate (reworded data) and analyze customer verbatims into demanded qualities (the WHATs) based on an understanding of how a customer plans to use a product or service. This table is composed of two parts. Part 1 contains information on *who, where, when* and *how* a product or service is or can be used by a customer. Part 2 sorts the reworded voices of the customers into possible categories for analysis, e.g., reliability, function, task and substitute quality characteristics.

Weight, absolute   Within the A1 matrix of the 30-matrix (GOAL/QPC) approach to QFD, the absolute weight is determined for each customer demand (the WHATs) by multiplying its customer priority/importance value by its improvement ratio by its sales point.

Weight, demanded   Within the A1 matrix of the 30-matrix (GOAL/QPC) approach to QFD, the demanded weight is the ratio of the absolute weight for each customer demand (the WHATs) to the sum of all the absolute weights.

WHAT–HOW concept   In the initial QFD matrix (the House of Quality or A1 matrix), the customer demands (the WHATs) are listed down the left side of the matrix and the substitute quality characteristics (the HOWs) are recorded across the top. The resulting L-shaped relationship matrix is used to fully delineate the extent to which each HOW is capable of satisfying or influencing each WHAT.

Z-series matrices   A collection of Software Quality Deployment (SQD) matrices created by Richard Zultner (see Chapter 11, Figures 11-1*a* and 11-1*b,* two views of the Z0, Z1 and Z2 matrices). These matrices are supplemental to the 30-matrix (GOAL/QPC) and 4-matrix American Supplier Institute approaches already used to assist in more traditional QFD projects on products and services.

BIBLIOGRAPHY

Adams, R. M. and M. D. Gavoor (1990). "Quality Function Deployment: Its Promise and Reality," *1990 ASQC Quality Congress Transactions,* San Francisco, CA: pp. 33–38.

Akao, Y. (1964). "Check Points, Control Points and Evaluation Points," *Quality Control,* JUSE, Vol. 15, Spring extra issue, pp. 42–48.

Akao, Y. (1969). "Quality-Featuring Characteristics of Quality Control," *Quality Control,* JUSE, Vol. 20, No. 5, pp. 37–41.

Akao, Y. (1972). "New Product Development and Quality Assurance-Quality Deployment System," *Standardization and Quality Control,* JSA, Vol. 25, No. 4, pp. 9–14.

Akao, Y., (1988a). *Practical Applications of Quality Deployment in New Product Development.* Japan Standards Association, Tokyo.

Akao, Y. (1988b). "A Series of Articles," presented at the GOAL/QPC Facilitator's Meeting, Aug. 1, 1988, GOAL/QPC, Methuen, MA.

Akao, Y., (1990a). "History of Quality Function Deployment in Japan," pp. 180–196 in Zeller, H. J., ed., *The Best on Quality: Targets, Improvements, Systems,* Vol. 3. New York: Hanser Publishers.

Akao, Y., T. Ohfuji and T. Naoi (1987). "Survey and Reviews on Quality Function Deployment in Japan," *ICQC—1987 Tokyo Proceedings,* B-1-02.

Akao, Y., S. Ono, A. Harada, H. Tanaka and K. Iwasawa (1983). "Quality Deployment including Cost, Reliability and Technology," *Quality,* JSQC, Vol. 13, No. 3, pp. 61–78.

Altshuller, G. *Creativity as an Exact Science.*

Anderson, D. M. (1990). *Design for Manufacturability, Optimizing Cost, Quality and Time-to-Market.* Lafayette, CA: CIM Press.

Anderson, D. M. "Design for Manufacturability," Anderson Seminars and Consulting, Lafayette, CA.

Anderson, D. M. "Designing in Quality and Reliability," Anderson Seminars and Consulting, Lafayette, CA.

Anderson, D. M. and B. J. Pine (1997). *Mass Customizing Products, The New Product Development Imperative,* Chicago: Irwin Professional Publishing.

Anthony, M. and A. Dirik (1995). "Simplified Quality Function Deployment for High-Technology Product Development," *Visions,* April, pp. 9–12.

ASI (1992). *Seven Management Planning Tools,* Allen Park, MI: American Supplier Institute.

Bahil, A. T. and W. L. Chapman (1993). "A Tutorial on Quality Function Deployment," *Engineering Management Journal,* Vol. 5, No. 3, pp. 24–35.

Barnard, W., R. Norman, W. Barnard and D. Daetz (1995). *Customer Integration: The Quality Function Deployment Leader's Guide for Decision Making.* New York: Wiley.

Berg, D. L. and W. M. Harral (1993). "Implementing TQM in an ISO Framework," pp. 153–159 in *1993 Annual Quality Congress Transactions,* sponsored by the American Society for Quality Control (ASQC). Boston, MA: ASQC.

Berg, D. L. and W. M. Harral (1995). "Closing the Loop," *Third Annual ISO 9000 Conference: Adding Value to Your Business Transactions,* sponsored by the American Society for Quality Control (ASQC). Washington, DC: ASQC.

Berglund, R. L. (1993). "QFD a Critical Tool for Environmental Decision Making," in *ASQC Quality Congress Transactions.* Boston, MA: American Society for Quality Control.

Bergman, B., A. Gustafsson and N. Gustafsson (1991). "Quality Function Deployment as a Tool for the Improvement of a Course in Total Quality Management and Methodology," presented at Second Renault-Volvo Symposium on Quality: Management and Advanced Techniques, Paris.

Bicknell, K. and B. Bicknell (1994). *The Road Map to Repeatable Success: Using QFD to Implement Change.* Cleveland, OH: CRC Press.

Blaine, R. C., D. W. Burden and N. E. Morrell (1989). "QFD: A Flexible Management Tool," in *Transactions of a Symposium on Quality Function Deployment, June 1989,* sponsored by ASQC, ASI and GOAL/QPC. Methuen, MA: GOAL/QPC.

Bralla, J. G. (1986). *Handbook of Product Design for Manufacturing: a Practical Guide to Low-Cost Design.* New York: McGraw-Hill.

Brassard, M. (1989). *Memory Jogger Plus +.* Methuen, MA: GOAL/QPC.

Brown, P. G. (1991). "QFD: Echoing the Voice of the Customer." *AT&T Technical Journal,* March/April.

Brown, S. W., E. Gummesson, B. Edvardsson and B. Gustavsson (1991). *Service Quality: Multidisciplinary and Multinational Perspectives.* Lexington, MA: Lexington Books.

Brunet, W. (1992). "ISO 9000 Series Standard Compared," *Quality Systems Update,* Vol. II, No. 8, p. 9.

Burrows, P. (1991). "In Search of the Perfect Product," *Electronic Business,* June 17, pp. 70–74.

Caron, J. R., S. L. Jarvenpaa and D. B. Stoddard (1994). "Business Reengineering at CIGNA Corp: Experiences and Lessons Learned from the First Five Years," *MIS Quarterly,* Sept., pp. 233–250.

Charteris, W. (1993). "Quality Function Deployment: A Quality Engineering Technology for the Food Industry," *Journal of the Society of Dairy Technology,* Vol. 46, No. 1, pp. 12–21.

Chen, C. and S. F. Bullington (1993). "Development of a Strategic Plan for an Academic Department Through the Use of Quality Function Deployment," *Computers and Industrial Engineering,* Vol. 25, Nos. 1–4, pp. 49–52.

Clausing, D. (1988). "Quality Function Deployment," in Ryan, N. E., ed., *Taguchi Methods and QFD.* Dearborn, MI: American Supplier Institute.

Clausing, D. (1994). *Total Quality Development,* New York: ASME Press.

Clausing, D. and S. Pugh (1991). "Enhanced Quality Function Deployment," presented at the Design and Productivity International Conference, 6–8, Feb. Honolulu, HI.

Cohen, L. (1988). "Quality Function Deployment: An Application Perspective from Digital Equipment Corporation," *National Productivity Review,* Vol. 7, No. 3, pp. 197–208.

Cohen, L. (1995). *Quality Function Deployment: How to Make QFD Work for You,* Reading, MA: Addison-Wesley.

Conti, T. (1989). "Process Management and Quality Function Deployment," *Quality Progress,* Dec., pp. 45–48.

Cox, C. A. (1992). "Keys to Success In Quality Function Deployment," *APICS—The Performance Advantage,* Vol. 2, No. 4, pp. 25–28.

Day, R. G. (1993). *Quality Function Deployment: Linking a Company with Its Customers.* Milwaukee, WI: ASQC Quality Press.

Dean, E. B. (1992). "Quality Function Deployment for Large Systems," in *Proceedings of the 1992 International Engineering Management Conference,* Oct. 25–28, Eatontown, NJ.

Deming, W. E. (1986). *Out of the Crisis.* Cambridge, MA: MIT Center for Advanced Engineering Study.

Made in America: Regaining the Productive Edge. by Dertouzos, M. L. R. K. Lester and R. M. Solow (1989). New York: Harper Perennial.

"Designing for Customer Satisfaction," *Management Decision,* Vol. 32, No. 5, 1994, pp. 37–38.

De Vera, D., T. Glennon, A. A. Kenny, M. A. H. Khan and M. Mayer (1988). "An Automotive Case Study," *Quality Progress,* June, pp. 35–38.

Ealey, L. (1987). "QFD—Bad Name for a Great System," *Automotive Industries,* July, p. 21.

Eccles, E. W. (1994). "Quality Function Deployment," *Engineering Designer,* Jan./Feb., pp. 9–11.

Ermer, D. S. (1995). "Using QFD Becomes an Educational Experience for Students and Faculty," *Quality Progress,* May, pp. 131–136.

Eureka, W. E. and N. E. Ryan (1994). *The Customer-Driven Company: Managerial Perspectives on Quality Function Deployment,* 2nd ed. Allen Park, MI: ASI Press.

Feedman, D. P. and G. M. Weinberg (1982). *Handbook of Walkthroughs, Inspections and Technical Reviews,* 3rd ed. Boston, MA: Little, Brown.

Feigenbaum (1961). *Total Quality Control,* 2nd ed.

Ferrell, S. F. and W. G. Ferrell, Jr. (1994). "Using Quality Function Deployment in Business Planning at a Small Appraisal Firm," *Appraisal, Journal,* Vol. 62, No. 3, pp. 382–390.

Florusse, L. B. and D. P. Clausing (1992). "A Detailed Framework for Quality Function Deployment Phase 3, Process Engineering," Working Paper, Massachusetts Institute of Technology, Cambridge, MA.

Fortuna, R. M. (1987). "Quality Function Deployment: Taking Quality Upstream," *Target* magazine, Association for Manufacturing Excellence, Wheeling, IL, Winter, pp. 11–16.

Frank, S. and J. Green (1992). "Applying Quality Function Deployment: A Team Approach to Design with QFD," *Army Research, Development, and Acquisition Bulletin,* May–June, pp. 14–19.

Gadd and Oakland (1996). "Chimera or Culture? BPR for TQM," *Quality Management Journal,* Fall.

Gane, C. and T. Sarson (1979). *Structured Systems Analysis: Tools and Techniques.* Englewood Cliffs, NJ: Prentice-Hall.

George, W. R. and B. E. Gibson (1991). "Blueprinting: A Tool for Managing Quality in Service," pp. 73–91 in Brown, S. W., E. Gummesson, B. Edvardsson and B. Gustavsson, eds. *Service Quality: Multidisciplinary and Multinational Perspectives.* Lexington, MA: Lexington Books.

GOAL/QPC Research Committee (1989). *Quality Function Deployment: A Process for Translating Customers' Needs into a Better Product and Profit.* Methuen, MA: GOAL/QPC.

GOAL/QPC Research Committee (1990). "Quality Function Deployment: A Process for Continuous Improvement," pp. 316–336 in *Transactions of the Second Symposium on Quality Function Deployment,* sponsored by ASQC, ASI and GOAL/QPC. Methuen, MA: GOAL/QPC.

Gopalakrishnan, K. N., B. E. McIntyre and J. C. Sprague (1992). "Implementing Internal Quality Improvement with the House of Quality," *Quality Progress,* Sept., pp. 57–60.

Graessel, R. and P. Zeidler (1993). "Using Quality Function Deployment to Improve Customer Service," *Quality Progress,* Nov., pp. 59–63.

Graham, R. E., R. B. Furnas and M. Babula (1993). *Design and Implementation of a Pilot Orientation Program for New NASA Engineering Employees,* May, NASA TM 105907.

Griffin, A. (1991). "Evaluating Development Processes: QFD as an Example," Report 91-121, Marketing and Management, University of Chicago, Chicago, IL.

Griffin, A. (1992). "Evaluating QFD's Use in US Firms as a Process for Developing Products," *Journal of Product Innovation Management,* Vol. 9, pp. 171–187.

Griffin, A. and J. R. Hauser (1992). "The Voice of the Customer," Working Paper, Report No. 92-106, Marketing Science Institute, Cambridge, MA, March.

Griffin, A. and J. R. Hauser (1993). "The Voice of the Customer," *Marketing Science,* Vol 12., No. 1, pp. 1–27.

Guinta, L. R. and N. C. Praizler (1993). The *QFD Book: The Team Approach to Solving Problems and Satisfying Customers Through Quality Function Deployment.* New York: American Management Association.

Gupta and Wileman (1990). "Accelerating the Development of Technology-Based New Products," *California Management Review,* Winter.

Gustafsson, N. (1995). "Comprehensive Quality Function Deployment—A Structured Approach for Design of Quality," Linköping Studies in Science and Technology Thesis No. 487, Linköping University, Linköping, Sweden.

Halbleib, L., P. Wormington, W. Cieslak and H. Street (1993). "Application of Quality Function Deployment to the Design of a Lithium Battery," *IEEE Transactions on Components, Hybrids, and Manufacturing Technology,* Vol. 16, No. 8, pp. 802–807.

Hales, R., D. Lyman, and R. Norman (1994). "QFD and the Expanded House of Quality," *Quality Digest,* Feb. 1994.

Hall, Rosenthal, and Wade (1993). "How to Make Engineering *Really* Work," *Harvard Business Review,* Nov./Dec. 1993.

Hammer, M. and J. Champy (1993). *Reengineering The Corporation: A Manifesto for Business Revolution.* New York: HarperBusiness.

Hatley, D. J. and I. A. Pirbhai (1987). *Strategies for Real-Time System Specification,* New York: Dorset House.

Hauser, J. R. (1993). "How Puritan-Bennett Used the House of Quality," *Sloan Management Review,* Vol. 34, No. 3. 1993.

Hauser, J. R. and D. Clausing (1988). "The House of Quality," *Harvard Business Review,* May-June, No. 3, pp. 63–73.

Havener, C. L. (1993). "Improving the Quality of Quality," *Quality Progress,* Nov., pp. 41–44.

Hicks, D. (1992). *Activity-Based Costing for Small and Mid-Sized Businesses.* New York: Wiley.

Hjort, H., D. Hananel and D. Lucas (1992). "Quality Function Deployment and Integrated Product Development," *Journal of Engineering Design,* Vol. 3, No. 1, pp. 17–29.

Hollins, W. and S. Pugh (1990). *Successful Product Design: What to Do and When.* London: Butterworth.

Hosotani, K. (1992). *Japanese Quality Concepts: An Overview,* transl. G. Mazur. Quality Resources: White Plains, NY.

Howe, D. R. (1983). *Data Analysis for Data Base Design.* Baltimore, MD: Edward Arnold.

Hrones, J. A. Jr., B. C. Jedrey, Jr. and D. Zaaf (1993). "Defining Global Requirements with Distributed QFD," *Digital Technical Journal,* Vol. 5, No. 4, pp. 36–46.

Hunter, M. R. and R. D. Van Landingham (1994). *Quality Progress,* April, pp. 55–59.

Imai, M. (1985). *Kaizen: The Key to Japan's Competitive Success.* New York: McGraw-Hill.

ISO 9000: Quality Management and Quality Assurance Standards—Guidelines for Selection and Use (1987a). Geneva, Switzerland: International Organization for Standardization.

ISO 9001: Quality System—Model for Quality Assurance in Design/Development, Production, Installation and Servicing (1987b). Geneva, Switzerland: International Organization for Standardization.

ISO 9002: Quality Systems—Model for Quality Assurance in Production and Installation (1987c). Geneva, Switzerland: International Organization for Standardization.

ISO 9003: Quality Systems—Model for Quality Assurance in Final Inspection and Test (1987d). Geneva, Switzerland: International Organization for Standardization.

ISO 9004: Quality Management and Quality System Elements—Guidelines (1987e). Geneva, Switzerland: International Organization for Standardization.

Ishikawa, K. (1985). *What is Total Quality Control? The Japanese Way.* Englewood Cliffs, NJ: Prentice-Hall.

Ishikawa, K. (1986). *Guide to Quality Control,* 2nd rev. ed. Tokyo: Asian Productivity Organization.

Jackson, H. K., Jr. and N. L. Frigon (1994). *Management 2000: The Practical Guide to World Class Competition.* New York: Van Norstrand Reinhold.

Jones, K. J. and T. Jackson (1996). *Implementing a Lean Management System.* Cambridge, MA: Productivity Press.

Kamizawa, N., I. Ishizuka and Y. Akao (1978). "Quality Evolution (Deployment) System and FMEA," *ICQC—1978 Tokyo Proceedings,* B4, pp. 19–28.

Kano, N., S. Nobuhiro, T. Fumio and T. Shinichi (1984a). "Attractive Quality and Must-be Quality," *Quality (Hinshitsu)*, Vol. 14 No. 2, pp. 39–48.

Kano, N., N. Seraku, F. Takahashi and S. Tsuji (1984b). "Attractive and Normal Quality," *Quality,* Vol. 14, No. 2, pp. 39–48.

Kawakita, J. 1986. *The KJ Method: Seeking Order Out of Chaos.* Tokyo: Chuokoron-sha (In Japanese).

King, R, (1987a). "Listening to the Voice of the Customer: Using the Quality Function Deployment System," *National Productivity Review,* Summer, pp. 277–281.

King, R. (1987b). *Better Designs in Half the Time: Implementing QFD in America.* Metheun, MA: GOAL/QPC.

King, R. (1987c). *Better Designs in Half the Time.* Methuen, MA: GOAL/QPC.

King, R. (1989a). *Better Designs in Half the Time: Implementing Quality Function Deployment in America.* Methuen, MA: GOAL/QPC.

King R. (1989b). "Affordable Innovation Using the Full Power of QFD's Matrix of Matrices to Get Timely, Cost Effective, Customer-Focused, Innovative Designs," pp. 71–125 in *Transaction of a Symposium on Quality Function Deployment, June 1989,* sponsored by ASQC, ASI and GOAL/QPC. Methuen, MA: GOAL/QPC.

King, R. (1995). *Designing Products and Services That Customers Want.* Methuen, MA: GOAL/QPC.

Kogure, M. and Y. Akao (1983). "Quality Function Deployment and CWQC in Japan: A Strategy for Assuring That Quality Is Built into New Products," *Quality Progress,* Oct., pp. 25–29.

Lam, K. D., F. D. Watson and S. R. Schmidt (1991). *Total Quality: A Textbook of Strategic Quality Leadership and Planning.* Colorado Springs, CO: Air Academy Press.

LaSala, K. (1994). "Identifying Profiling System Requirements with Quality Function Deployment," pp. 249–254 in *Proceedings of the Fourth Annual International Symposium of the National Council on Systems Engineering,* Vol. 1., Aug. 10–125. San Jose, CA.

Lee, K. (1995). "A Method to Incorporate Optimization and Fuzzy Information in Quality Function Deployment," Ph.D. Dissertation, College of Engineering, Wichita State University, Wichita, KS.

Lyman, D., R. F. Buesinger and J. P. Keating (1994). "QFD in Strategic Planning," *Quality Digest,* May, pp. 45–52.

Lu, M. H., C. N. Madu, C. Kuei and D. Winokur (1994). "Integrating QFD, AHP and Benchmarketing in Strategic Marketing," *Journal of Business and Industrial Marketing,* Vol. 9, No. 1, pp. 41–50.

McElroy, J. (1987). "The House of Quality: For Whom Are We Building Cars," *Automotive Industries,* June, pp. 68–70.

McElroy, J. (1989). "QFD: Building the House of Quality," *Automotive Industries,* Jan., pp. 30–32.

McMenamin, S. M. and J. F. Palmer (1984). *Essential Systems Analysis.* New York: Yourdon Press.

Maddux, G. A., R. W. Amos and A. R. Wyskida (1991). "Organizations Can Apply Quality Function Deployment as Strategic Planning Tool," *Industrial Engineering,* Sept., pp. 33–37.

Maduri, O. (1992). "Understanding and Applying QFD in Heavy Industry," *Journal for Quality and Participation,* Jan./Feb., pp. 64–69.

Mallon, J. C. and D. E. Mulligan (1993). Quality Function Deployment—A System for Meeting Customers' Needs," *Journal of Construction Engineering and Management,* Vol. 119, No. 3, pp. 516–531.

Malsbury, J. A. (1987). "Why Johnny Can't Program," pp. 769–773 in *Forty-First Annual Quality Congress Transactions.* Milwaukee, WI: American Society for Quality Control.

Marsh, S., J. W. Moran, S. Nakui and G. Hoffherr (1991). *Facilitating and Training in Quality Function Deployment.* Methuen, MA: GOAL/QPC.

Masud, A. S. M. and E. B. Dean (1993). "Using Fuzzy Sets in Quality Function Deployment," in *Proceedings of the Second Industrial Engineering Research Conference,* May 26–27, Los Angeles CA.

Mazur, G. (1991). "Voice of the Customer Analysis and Other Recent QFD Technology," *Transactions from the Third Symposium on Quality Function Deployment,* June, pp. 285–297, Dearborne, MI.

Mazur, G. (1992). "Task Deployment for Service QFD and the Americans with disabilities Act," *Proceedings from the GOAL/QPC Ninth Annual Conference,* Boston, MA.

Mehta, P. (1994). "Designed Chip Embeds User Concerns," *Electronic Engineering Times,* Jan. 24.

Mizuno, S. ed. (1988). *Management for Quality Improvement: The 7 New QC Tools.* Cambridge, MA: Productivity Press.

Mizuno, S. and Y. Akao (1978). *Quality Function Deployment: An Approach to Company-wide Quality Control.* Tokyo: Japanese Union of Scientists and Engineers (JUSE).

Mizuno, S. and Y. Akao, ed. (1994). *QFD: The Customer-Driven Approach to Quality Planning and Development.* Tokyo: Asian Productivity Organization (APO); also available from Quality Resources Press, New York City.

Mizuno, S. and Y. Akao, ed. (1993). *Quality Function Deployment,* rev. ed., transl. G. Mazur. White Plains, NY: Quality Resources.

Moder, J. J., C. R. Phillips and E. W. Davis (1983). *Project Management with CPM, PERT and Precedence Diagramming,* 3rd ed. New York: Van Norstrand Reinhold.

Moffat, S. (1990). "Japan's New Personalized Production," *Fortune,* Oct. 22, p. 132.

Moran, J. W. and J. B. ReVelle (1994). *The Executive's Handbook on Quality Function Deployment: Defining Management's Roles and Responsibilities.* Windham, NE: Markon.

Morrell, N. E. (1988). "Quality Function Deployment," Society of Automotive Engineers Technical Paper Series #870272, Warrendale, PA.

"Motown's Struggle to Shift on the Fly," *Business Week,* July 11, 1994, p. 111.

Mouradian, G. (1992). "A Defense Contractor's View of ISO 9000," in *Proceedings of an ISO 9000 Quality Standards Clinic,* Sept. 10, 1992, Chicago, IL, sponsored by the Society of Manufacturing Engineers. Dearborn, MI: Society of Manufacturing Engineers.

Murgatroyd, S. (1993). "The House of Quality: Using QFD for Instructional Design in Distance Education," *American Journal of Distance Education,* Vol. 7, No. 2, pp. 34–48.

Myers, G. J. (1979). *The Art of Software Testing.* New York: Wiley.

Nakui, S. (1991). "Comprehensive QFD System," *Transactions from the Third Symposium on Quality Function Deployment,* June, 1991, Dearborne, MI.

Nakui, S. and J. Terninko (1992a). "Structuring a Quality Design Process Chart," presented at GOAL/QPC 9th Annual Conference, Boston, MA. 1992.

Nakui, S. and J. Terninko (1992b). "A Road Map to a Better Product Design Process: Structuring a Quality Design Process Chart (QDPC)," in *Proceedings from the GOAL/QPC Ninth Annual Conference.* Methuen, MA: GOAL/QPC.

Nelson, D. (1992). "The Customer Process Table: Hearing Customers' Voices Even If They're Not Talking," *Transactions from the Fourth Symposium on Quality Function Deployment,* June.

"News Backgrounder: TQM," *ASQC News Bureau Press Release.* Milwaukee: ASQC, May 1991.

Nishimura, K. (1972). "Ship Design and Quality Table," *Quality Control,* JUSE, Vol. 23, May extra issue, pp. 16–20.

Norman, R., B. Dacey and D. Lyman (1991). "QFD: A Practical Implementation,: *Quality,* Vol. 30, No. 5, pp. 36–40.

Ohfuji, T., Mi Ono and Y. Akao (1990). *Quality Deployment Methods: Procedures and Practices for Making Quality Charts.* Tokyo: Union of Japanese Scientists and Engineers [in Japanese].

O'Neal, C. R. and W. C. LaFief (1992). "Marketing's Lead Role in Total Quality," *Industrial Marketing Management,* Vol. 21, pp. 133–143.

Ono, M. and T. Ohfuji (1990). *Expressing Demanded Quality on Quality Function Deployment.* Tokyo: Japanese Society for Quality Control.

Ozeki, K. and T. Asaka (1990). *Handbook of Quality Tools: The Japanese Approach.* Cambridge, MA: Productivity Press.

Pardee, W. (1996). *To Satisfy and Delight Your Customer: How to Manage for Customer Value.* Dorset House, 1991.

Peters, T. (1992). *Liberation Management: Necessary Disorganization for the Nanosecond Nineties.* New York: Alfred A. Knopf.

Phadke, M. (1989). *Quality Engineering Using Robust Design,* Englewood Cliffs, NJ: Prentice-Hall.

Pine II, B. J. (1993). *Mass Customization: The New Frontier in Business Competition.* Boston: Harvard Business School Press.

Proceedings from the First Pacific Rim Symposium on Quality Deployment, Macquarie Graduate School of Management, Macquarie University, Sydney, Feb. 15–17, 1995.

Proceedings from the Symposium on Integrated Product and Process Design, ASI, Allen Park, MI, 1995.

Pugh, S. (1981). "Concept Selection—A Method That Works," International Conference on Engineering Design ICED 81, Rome, Italy, March 9–13. March.

Pugh S. (1991). *Total Design Integrate Methods for Successful Product Engineering.* Reading, MA: Addison-Wesley.

Quality Function Deployment: A Collection of Presentations and QFD Case Studies (1987). Dearborn, MI: American Supplier Institute.

Quality Function Deployment Awareness Seminar (1989). Dearborn, MI: Quality Education and Training Center, Ford Motors.

Quality Measure, The (*Le Qualimètre*). Mouvement Québécois de la Qualité, 455, rue Saint-Antonine Quest , bureau 404, Montreal, Québec, H2Z 1J1, Canada, 1995.

Rappaport, A. (1986). *Creating Shareholder Value.* New York: Free Press.

Raynor, M. E. (1994). "The ABC's of QFD: Formalizing the Quest for Cost-Effective Customer Delight," *National Productivity Review,* Summer, pp. 351–357.

Reed, B. M. and D. A. Jacobs (1993). "Guidelines for Implementation of Quality Function Deployment (QFD) in Large Space Systems," Final Report, December, NAS1-19859 (Task 28).

Reed, B. M., D. A. Jacobs and E. B. Dean (1994). "Quality Function Deployment: Implementation Considerations for the Engineering Manager," pp. 2–6 in *Proceedings of the IEEE International Engineering Management Conference,* Oct. 17–19, Dayton, OH.

ReVelle, J. B. (1988). *The New Quality Technology.* Los Angeles, CA: Hughes Aircraft Co.

ReVelle, J. B., N. L. Frigon, Sr. and H. K. Jackson, Jr. (1995). *From Concept to Customer: The Practical Guide to Integrated Product and Process Development, and Business Process Reengineering.* New York: Van Nostrand Reinhold.

Reynolds, M. J. (1989). "Industry Professionals Speak Out? The Importance of the Quality Concept," *Elastomerics,* Nov., pp. 13–16.

Ryan, N. E., ed. (1988). *Taguchi Methods and Quality Function Deployment: Hows and Whys for Management.* Allen Park, MI: ASI Press.

Ryan, N. E. and W. Eureka (1994). Quality Up, Costs Down: A Manager's Guide to Taguchi Methods and QFD. Homewood, IL: Irwin.

Saaty, T. L. (1988). *Decision Making for Leaders: The Analytic Hierarchy Process for Decisions in a Complex World.* Pittsburgh, PA: RWS Publications.

Saaty, T. L. (1990). *Decision Making for Leaders: The Analytic Hierarchy Process for Decisions in a Complex World,* rev. 2nd ed. Pittsburgh, PA: RWS Publications.

Sawada, T. (1979). "Quality Deployment of Product Planning," pp. 27–30 in *Proceedings of the Ninth Annual Convention of Japan Society for Quality Control.*

Schaal, H. F. and W. R. Slabey (1990). *Implementing QFD at the Ford Motor Company.* Dearborn, MI: Ford Motor Company.

Schauerman, S., D. Manno and B. Peachy (1994). "Listening to the Customer: Implementing Quality Function Deployment," *Community College Journal of Research and Practice,* Vol. 18, pp. 397–409.

Schonberger, R. J. (1986). *World Class Manufacturing: The Lessons of Simplicity Revealed.* New York: Free Press.

Schubert, M. A. (1989). "Quality Function Deployment—A Means of Integrating Reliability Throughout Development," pp. 93–98 in *Proceedings of the Society of American Value Engineers Conference.*

Schuldt, G. A. (1986). "An Introduction to Access Modeling," in *SDF VII Conference Proceedings.* San Francisco, CA: Structured Development Forum.

Shillito, M. L. (1992). "Quality Function Deployment: The Total Product Concept," in Shillito, M. L. and D. J. De Marle, eds., *Value, Its Measurement, Design, and Management.* New York: Wiley.

Shillito, M. L. (1994). *Advanced QFD: Linking Technology to Market and Company Needs.* New York: Wiley.

Smith, L. (1989). *QFD and Its Application in Concurrent Engineering,* Slabey, W. R., ed. Ford Motor Company.

Snodgrass, T. J. and M. Kasi (1986). *Function Analysis: The Stepping Stones to Good Value.* Madison, WI: University of Wisconsin.

Stocker, G. D. (1991). "Using QFD to Identify Customer Needs," *Quality Progress,* Jan., p. 120.

Sullivan, L. P. (1986). "Quality Function Deployment—A System to Assure that Customer Needs Drive the Product Design and Production Process," *Quality Progress,* June, pp. 39–50.

Terninko, J. (1989). *Robust Design: Key Points for World Class Quality.* Nottingham, NH: Responsible Management.

Terninko, J. (1991). "QFD Synergy with Taguchi's Philosophy," presented at QOAL/QPC Eighth Annual Conference, Boston, MA.

Tomita, Y., K. Ishikawa, K. Kondo and K. Watanuki (1969). "An Example of Quality Assurance Activities," *ICQC-1969 Tokyo Proceedings,* pp. 243–246.

Tone, K. and R. Manabe, ed. (1990). *Case Studies in AHP.* Tokyo: JUSE Press [in Japanese].

Tool and Manufacturing Engineers Handbook, Vol. 6: *Design for Manufacturability* (1992). Dearborn, MI: Society of Manufacturing Engineers (SME).

TQM Committee Report, ASQC Quality Management Division, Feb. 1993.

Transactions from A Symposium on Quality Function Deployment, June 19–20, Novi, MI, Automotive Div.-ASQC, Milwaukee, WI, American Supplier Institute, Dearborn, MI, and GOAL/QPC, Methuen, MA (1989) [all transactions listed and categorized in Appendix B].

Transactions from the Second Symposium on Quality Function Deployment, June 18–19, Novi, MI, GOAL/QPC, Methuen, MA, Automotive Div., ASQC, Milwaukee, WI, and American Supplier Institute, Dearborn, MI (1990) [all transactions listed and categorized in Appendix B].

Transactions from the Third Symposium on Quality Function Deployment, June 24–25, Novi, MI, American Supplier Institute, Dearborn, MI, and GOAL/QPC, Methuen, MA (1991) [all transactions listed and categorized in Appendix B].

Transactions from the Fourth Symposium on Quality Function Deployment, June 15–16, Novi, MI, GOAL/QPC, Methuen, MA, and American Supplier Institute, Dearborn, MI (1992) [all transactions listed and categorized in Appendix B].

Transactions from the Fifth Symposium on Quality Function Deployment, June 21–22, Novi, MI, American Supplier Institute, Dearborn, MI, and GOAL/QPC, Methuene, MA (1993). [all transactions listed and categorized in Appendix B].

Transactions from the Sixth Symposium on Quality Function Deployment, June, Novi, MI, GOAL/QPC, Methuen, MA, American Supplier Institute, Allen Park, MI, and QFD Institute, Ann Arbor, MI (1994) [all transactions listed and categorized in Appendix B].

Transactions from the Seventh Symposium on Quality Function Deployment, June 11–13, Novi, MI, QFD Institute, Ann Arbor, MI (1995) [all transactions listed and categorized in Appendix B].

Transactions from the Eighth Symposium on Quality Function Deployment, June 9–11, Novi, MI, QFD Institute, Ann Arbor, MI (1996) [all transactions listed and categorized in Appendix B].

Trucks, H. E. (1987). *Designing for Economical Production,* 2nd ed. Dearborn, MI: Society of Manufacturing Engineers (SME).

Ungvari, S., "Total Quality Management and Quality Function Deployment," pp. 103–135 in *Third Symposium on Quality Function Deployment, June 1991,* sponsored by ASI and GOAL/QPC. Methuen, MA: GOAL/QPC.

Urban, G. and J. Hauser (1993). *Design and Marketing of New Products,* 2nd ed. Englewood Cliffs, NJ: Prentice-Hall.

Van Treeck, G., and R. Thackeray (1991). "Quality Function Deployment at Digital Equipment Corp.," *Concurrent Engineering,* Vol. 1, No. 1, Jan./Feb.

Warfield, J. (1994). *A Science of Generic Design: Managing Complexity Through System Design,* 2nd ed. Iowa State University Press, Ames, Iowa.

Wasserman, G. S. (1993). "On How to Prioritize Design Requirements During the QFD Planning Phase," *IEE Transactions,* Vol. 25, No. 3, pp. 59–65.

"What's QFD? Quality Function Deployment Quietly Celebrates Its First Decade in the U.S.," *Industry Week,* Nov. 1, 1993.

Wheelwright, S. and K. Clark (1995). *The Product Development Challenge: Competing Through Speed, Quality and Creativity.* Cambridge, MA: Harvard Business School Press.

White, C. (1966). *Quality Digest,* July.

Womack, J. P., D. T. Jones, and D. Roos (1990). *The Machine That Changed the World: The Story of Lean Production.* New York: Harper Perennial.

Zachman, J. A. (1986). "Framework for Information Systems Architecture," in *SDF VII Conference Proceedings.* San Francisco, CA: Structured Development Forum.

Zultner, R. E. (1988). "The Deming Approach to Software Quality Engineering," *Quality Progress,* Vol. 21, No. 11, pp. 58–64.

Zultner, R. E. (1989). "Objectives: The Missing Piece of the Requirements Puzzle," in *STA 5 Conference Proceedings.* Chicago, IL: Structured Techniques Association.

Zultner, R. (1992). "Task Deployment for Service," in *Transactions from the Fourth Symposium on Quality Function Deployment,* June, Dearborne, MI.

VIDEOS

Total Quality Management—Phase I: Quality Function Deployment (TQM with Jack B. ReVelle series), The George Washington University CEEP Books & Videos, 2020 K Street NW, Suite 240, Washington, DC 20052.

INDEX